THE PHYSICS OF
BASEBALL

2nd Edition, Revised
ROBERT K. ADAIR

Blending scientific facts and what a baseball does when tions—and why. This completely revised edition of *The Physics of Baseball* contains ne~ ~g, long home runs, shagging flie ~ ~ Denver (home of the Colorado Rockies) affect batted and pitched balls. Complete with understandable charts and graphs, baseball lore, and entertaining anecdotes about various players and incidents, *The Physics of Baseball* will delight and entertain baseball fans and physics enthusiasts and provide a whole new perspective on the game.

Robert K. Adair is Sterling Professor of Physics at Yale University and a member of the National Academy of Sciences whose research has largely been concerned with the properties of the elementary particles and forces of the universe.

"The Physics of Baseball addresses mysteries about the physical laws of the game that serious fans have pondered long enough for the members of the hot stove league to burn low."
—*New York Times*

"A brilliant book . . . written in a clear, elegant style."
—A. Bartlett Giamatti

THE PHYSICS OF BASEBALL

PHYSICS

BASEBALL

tion, Revised, Updated, and Enlarged

ROBERT KEMP ADAIR

Sterling Professor of Physics
Yale University
and
Physicist to the National League 1987–1989

HarperPerennial
A Division of HarperCollins*Publishers*

HarperCollins books may be purchased for educational, business, or sales promotional use. For information please write: Special Markets Department, HarperCollins Publishers, Inc., 10 East 53rd Street, New York, NY 10022.

FIRST EDITION

LIBRARY OF CONGRESS CATALOG CARD NUMBER 89-45623

ISBN 0-06-095047-1

94 95 RRD(H) 10 9 8 7 6 5 4

To the memory of my grandfather, Theodore Wiegman (1876–1953), who often sat with a small boy behind third base watching 3-I League baseball in the old Fort Wayne ball park in the 1930s, and to the memory of my son, James Cleland Adair (1957–1978), graceful first baseman on his championship Little League team.

CONTENTS

LIST OF FIGURES

PREFACE

Late in the summer of 1987, Bart Giamatti, then President of the National Baseball League and later Commissioner of Baseball, an old friend and colleague of mine from his days as Professor of English and then President of Yale, asked me to advise him on the (few) elements of baseball that might be best addressed by a physicist. I told Bart that I would be delighted to do so—and that I expected I would have so much fun at such a job that I would find it incorrect to accept any payment as consultant. Bart, ever the English professor and sensitive to words, responded by appointing me "Physicist to the National League"—a title that absolutely charmed the ten-year-old boy who I hope will always be a part of me.

When I considered the few questions Bart posed for me, it became increasingly clear that to answer any question properly, I had better understand almost all baseball as best I could. Hence, more as a delightful hobby than as an obligation, I attempted to describe quantitatively as much of the action of baseball as I could in a report to Bart Giamatti as League President. Bart then suggested that I publish the report, suitably expanded, as a book. I was pleased that Bart saw the final manuscript before his death and that he liked it.

Written for fun, and originally for Bart Giamatti, this book is not meant as a scholarly compendium of research on baseball—though I have borrowed extensively from the work of others. Moreover, it is not meant as the definitive treatise on the aspects of baseball that are considered. I have not hesitated to make best estimates on matters that I do not understand as well as I would like, and, though I hope I have made no egregious blunders, the physics of baseball is not trivial and I may have slipped somewhere.

Although I found no errors in the first edition central to baseball, there were numerous less significant infelicities and I have made a number of modest changes in numbers and text in this second edition to reflect my improved understanding of the sport, and I have added a sprinkling of comments on small points that I had not previously addressed. These changes modify, but do not radically change, the analyses found in the first edition.

Aside from such modest corrections and additions, I have also added some substantial material on matters that I found of interest to myself, fans, and players. I have added a chapter on running the bases, fielding fly balls, and throwing from the infield and outfield, and I have added sections on hitting left- or right-handed, the hitting of long home runs, and on baseball played at high altitudes—in particular, at the altitude of Denver.

Designed for those interested in baseball, not in the simple principles of physics, I have slighted descriptions of the details of the calculations to which I refer. These calculations were usually conducted by simple BASIC programs on a personal computer. The few formulas which are included, segregated as Technical Notes at the end of each chapter, are introduced only to provide a succinct description of the models I used for those who might be especially interested.

Robert Kemp Adair
Hamden, Connecticut

THE PHYSICS OF BASEBALL

MODELS AND THEIR LIMITATIONS

A small, but interesting, portion of baseball can be understood on the basis of physical principles. The flight of balls, the liveliness of balls, the structure of bats, and the character of the collision of balls and bats are a natural province of physics and physicists.

In his analysis of a real system, a physicist constructs a well-defined model of the system and addresses the model. The system we address here is baseball. In view of the successes of physical analyses in understanding arcane features of nature—such as the properties of the elementary particles and fundamental forces that define our universe (my own field of research) and the character of that universe in the first few minutes of creation—it may seem curious that the physics of baseball is not at all under control. We cannot calculate from first principles the character of the collision of an ash bat with a sphere made up of layers of different tightly wound yarns, nor do we have any precise understanding of the effect of the airstream on the flight of that sphere with its curious yin-yang pattern of stitches. What we can do is construct plausible models of those interactions that play a part in baseball which do not violate basic principles of mechanics. Though these basic principles—such as the laws of the conservation of energy and momentum—severely constrain such models, they do not completely define them. It is necessary for the models to touch the results of observations—or the results of the controlled observations called experiments—at some points so that the model can be more precisely defined and used to interpolate known results or to extrapolate such results. Baseball, albeit rich in anecdote, has not been subject to extensive quantitative studies of its mechanics, hence, models of baseball are not as well founded as they might be.[1]

However connected with experience, model and system—map and territory—are not the same. The physicist can usually reach precise conclusions about the character of the model. If the model is well chosen so as to represent the salient points of the real system adequately, conclusions derived from an analysis of the model can apply to the system to a useful degree. Conversely, conclusions—although drawn in a logically impeccable manner from premises defined precisely by the model—may not apply to the system because the model is a poor map of the system.

Hence, in order to consider the physics of baseball, I had to construct an ideal baseball game which I could analyze that would be sufficiently close to the real game so that the results of the analysis would be useful. The analysis was easy; the modeling was not. I found that neither my experience playing baseball (poorly) as a youth nor my observations of play by those better fitted for the game than I ideally prepared me for the task of constructing an adequate model of the game. However, with the aid of seminal work by physicists Lyman Briggs, Paul Kirkpatrick,[2] and others, and with help from discussions with other students of the game,

---

[1]Curiously, better analyses have been made of golf, probably because there are economic advantages to the support of research by manufacturers who might make and sell better balls and better clubs. Baseballs and (largely) baseball bats are made to specifications set down by major league officials and are less subject to manufacturers' improvements.

[2]Papers by Briggs and Kirkpatrick were published in the *American Journal of Physics*, Briggs's paper in 1959 and Kirkpatrick's paper in 1963. There is some scientific literature on the physics of baseball. As with most fields of science, some of this work is wrong. Science is difficult, and original published results—the raw materials of science—as yet untouched by the sifting and winnowing process that results in reliable knowledge, are not always valid. I try to report sifted results and give the reader some idea of uncertainties as I see them.

such as my long-time associate R. C. Larsen, I believe I have been able to arrive at a sufficient understanding of baseball so that some interesting conclusions from analyses of my construction of the game are relevant to real baseball.

In all sports analyses, it is important for a scientist to avoid hubris and pay careful attention to the athletes. Major league players are serious people, who are intelligent and knowledgeable about their livelihood. Specific, operational conclusions held by a consensus of players are seldom wrong, though—since baseball players are athletes, not engineers or physicists—their analyses and rationale may be imperfect. If players think they hit better after illegally drilling a hole in their bat and filling it with cork, they must be taken seriously. The reasons they give for their "improvement," however, may not be valid. I hope that nothing in the following material will be seen by a competent player of the game to be definitely contrary to his experience in playing the game. Honed by a century of intelligent trial and error, baseball must surely be played correctly—though not everything *said* about the play, by players and others, is impeccable. Hence, if a contradiction arises concerning some aspect of my analyses and the way the game is actually played, I would presume it likely that I have either misunderstood that aspect myself or that my description of my conclusion was inadequate and subject to misunderstanding.

Even as the results discussed here follow from analyses of models that can only approximate reality, the various conclusions have different degrees of reliability. Some results are quite reliable; the cork, rubber, or whatever stuffed into holes drilled in bats certainly does not increase the distance the ball can go when hit by the bat. Some results are hardly better than carefully considered guesses: How much does backspin affect the distance a long fly ball travels? Although I have tried to convey the degree of reliability of different conclusions, it may be difficult to evaluate the caveats properly. By and large, the qualitative results are usually reliable, but most of the quantitative results should be considered with some reserve, perhaps as best estimates.

In spite of their uncertainties, judiciously considered quantitative estimates are interesting and important; whatever their uncertainties, they often supplant much weaker—and sometimes erroneous—qualitative insights. Consequently, I have attempted to provide numerical values almost everywhere: sometimes when the results are somewhat uncertain, sometimes when the numbers are quite trivial but not necessarily immediately accessible to the reader.

As this exposition is directed toward those interested in baseball, not physics, I have chosen to present quantitative matters in terms of familiar units using the English system of measures—distances in feet and inches, velocities in miles per hour (mph), and forces in terms of ounce and pound weights. I have also often chosen to express effects on the velocities of batted balls in terms of deviations of the length of a ball batted 400 feet (a long home run?) under standard conditions.

To express the goals of this book, I can do no better than adopt a modification of a statement from Paul Kirkpatrick's article "Batting the Ball": *The aim of this study is not to reform baseball but to understand it.* As a corollary to the statement of purpose, I must emphasize that the book is not meant as a guide to a player; of all the ways to learn to throw and bat a ball better, an academic study of the mechanics of the actions must be the least useful.

THE FLIGHT OF THE BASEBALL

THE BASEBALL—AIR RESISTANCE

From the *Official Baseball Rules:*

> 1.09 The ball should be a sphere formed by yarn wound around a small sphere of cork, rubber, or similar material covered with two stripes of white horsehide or cowhide, tightly stitched together. It shall weigh not less than 5 nor more than $5\frac{1}{4}$ ounces avoirdupois and measure no less than 9 nor more than $9\frac{1}{4}$ inches in circumference.

The description in the rule book, ingenuous and charming, is not that of an engineer; the manufacturer (once in Chicopee, Massachusetts, then Haiti, then Taiwan, and, in 1989, Haiti again) requires further directions: "The cork nucleus, enclosed in rubber, is wound with 121 yards of blue-gray wool yarn, 45 yards of white wool yarn, and 150 yards of fine cotton yarn. Core and winding are enclosed by rubber cement and a two piece cowhide [horsehide before 1974] cover hand-stitched together with just 216 raised red cotton stitches."

Much more is required to define completely the ball that is the center of the sport of baseball, but its flight is largely set by the size and weight constraints listed in the rules. The paths of baseballs projected at velocities common to the game are strongly influenced by air resistance. The forces on the ball from this resistance are typically of the same magnitude as the force of gravity. A ball batted with an initial velocity of 110 mph at an angle of 35° from the horizontal would go about 750 feet in a vacuum; at Shea Stadium in New York, it will travel only about 400 feet. Hence, it is necessary to understand the fluid dynamics of air flow about spheres to understand the flight of the baseball.

When an object (such as a baseball) passes through a fluid (such as air), the fluid affects the motion of the object as it flows about the object. Moreover, for all fluids and all objects, the character of the flow of the fluid is determined by the value of a (dimensionless) Reynolds number proportional to the density of the fluid, the fluid velocity, the size of the object, and inversely proportional to the viscosity of the fluid[a]. For a given Reynolds number, the behavior of the gaseous fluid of stars—interacting with each other through gravity—that make up a galaxy 100,000 light-years across is described in very much the same way as the behavior of the molecules of air passing through an orifice one micron across, where a micron is about equal to the resolution of a high-power microscope.

The most interesting actions in the game of baseball take place when velocities of the ball range from a few mph (and Reynolds numbers of 10,000) to values near 120 mph (and Reynolds numbers near 200,000). For velocities in that range below about 50 mph, the flow of the air about the ball is rather smooth, though trailing vortices are generated. This air flow does not actually reach the surface of the ball where there is a quiet (Prandtl) boundary layer. A very, very small insect (perhaps a plant aphid) sitting on the moving ball would feel no breeze at all. At velocities above 200 mph the flow penetrates the boundary layer (the aphid would have to hold on very tight if it were not to be blown off), and the air at the boundary—and trailing behind the ball—is quite turbulent. We label the two regions conveniently (though a little inaccurately) as *smooth* and *turbulent.*

Thus, for a baseball passing through air at a velocity less than 50 mph the air flow is smooth, but the air flow is turbulent for velocities greater than 200 mph. Much of the subtlety of baseball is derived from the fact that so much of the game is played in the region between definitely smooth flow and definitely turbulent flow, at ball velocities greater than 50 mph and smaller than 120 mph. For balls traveling at the transition velocities between 50 mph and 120 mph, the flow can be smooth or turbulent depending on the detailed character of the surface of the ball and its motion. By and large, turbulence will be induced at lower velocities by roughness in the surface—and held off at higher velocities if the surface is very smooth. Furthermore, at a given velocity, the air resistance is, surprisingly, *smaller* for turbulent flow than for smooth flow.

From our understanding of fluid flow, it is convenient to describe the force on a moving baseball[b] (or equivalent sphere) as proportional to the cross-sectional area of the ball, proportional to the square of the velocity of the ball (double the velocity, increase the drag by a factor of four), proportional to the density of the air, and proportional to a quantity called the drag coefficient, which varies slowly with the velocity of the ball. Since the size of the ball is fixed and the density of the air does not vary much for the conditions under which baseball is played, we can consider that the force on the ball is proportional to the square of the velocity of the ball and the drag coefficient, which depends only on the velocity through the value of the Reynolds number.

The solid line on the graph of Figure 2.1 shows an estimate of the variation of the drag coefficient for a baseball as a function of the velocity of the ball. The drag will also depend to some extent on the orientation of the stitches on the ball. When the ball is rotating—as is usually the case—the drag will depend on the position of the axis of rotation with respect to the body (i.e., the stitches) of the ball, on the direction of the axis with respect

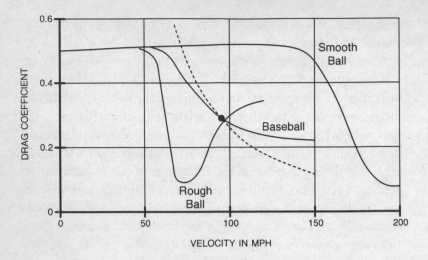

FIGURE 2.1 *The drag coefficient vs. ball velocity for various balls the size of a baseball. At the broken line the force on the ball equals gravity.*

to the ground and the direction of the ball's flight, and on the velocity of rotation of the ball. The drag on a rapidly spinning ball is probably slightly greater than that on a slowly rotating ball. But that effect is probably small; we estimate that the incremental drag on the spinning ball will usually not be much greater than five percent of the drag on the nonspinning ball; for a 90-mph fast ball thrown with a spin of 1500 rpm, the extra drag will reduce the speed of the ball as it crosses the plate no more than 0.5 mph.

For rotating balls, the dependence of the drag on these factors is not likely to be large, however, and we can consider that the drag effects described here represent a kind of average over configurations and spins. Aside from these caveats, the values of the coefficients for a baseball are not well known, but wind tunnel measurements have been made of the force on a ball at velocities up to 95 mph that support the solid curve. In particular, wind tunnel measurements showed the ball suspended nearly motionless in an upward-directed 95-mph vertical airstream. Therefore, for a ball moving through the air with a velocity of 95 mph, the drag force is about equal to the weight of the ball.

The broken curved line in the figure corresponds to values of the drag coefficient that would generate a drag equal to the force of gravity. The ball suspended in the airstream is held steady by the force of the air. This is the case only if the force is greater if the ball is falling and hence moving faster through the air and if the force is weaker if the ball is rising and moving slower with respect to the air. Consequently, the observed stability of the ball demands that the solid curve cross the broken line from left to right, further defining the variation of the drag coefficient with the velocity of the baseball.

The values expressed for larger velocities up to 150 mph are estimates, albeit guided by theoretical considerations. The mean uncertainties are perhaps 10 percent for velocities less than 120 mph, which is near the highest velocity reached by a ball in play. The rather gradual reduction in the drag coefficient from the value of about 0.5 for velocities less than 50 mph to the values of about 0.2 for velocities greater than 120 mph suggests that the transition from smooth flow to turbulent flow of the air passing the baseball in flight occurs gradually.

The values of the drag coefficient for an ideally smooth ball and a uniformly rough ball—about as rough as a ball completely covered with stitches—are shown also.[1] The variation of the drag coefficient with velocity will have the same general character for a ball that is—uniformly—a little less, or a little more, rough, but the drag minimum will be found at lower velocities for a rougher ball and at higher velocities for a less rough—or smoother—ball.

At the velocities of 50 to 130 mph dominant in baseball, the air passes over a smooth ball the size of a baseball in a smooth high-resistance flow. Turbulence is not induced until velocities near 200 mph. A real baseball, however, with its raised stitch

---

[1] These properties of balls of various degrees of roughness were established experimentally by E. Aschenbach and published in 1974 in the *Journal of Fluid Mechanics*.

patterns, induces low-resistance turbulent flow at the baseball velocities. Consequently, if the baseball were quite smooth rather than provided with protuberant stitches, as is the case, it could not be thrown or batted nearly as far—a stitched baseball batted 400 feet would travel only about 300 feet if it were very smooth. This effect is dramatic in golf; the air resistance from the smooth flow about a smooth ball would be so great that the ball would go nowhere. The ball is thus artificially roughened by the dimples impressed in the covering to induce turbulence and reduce the air resistance.

At velocities near 175 mph, where the resistive force for smooth balls falls off sharply, the resistance on the ball actually becomes smaller as the velocity of the ball increases. The resistance on a smooth ball the size of a baseball traveling 190 mph is smaller than the resistance on such a ball traveling 160 mph. This sharp dip in the drag coefficient at the onset of turbulence has been called the "drag crisis." For uniformly rough balls the sharp reduction in drag with increasing velocity comes at lower velocities; the rougher the ball the lower the velocity at which turbulence is induced and the lower the velocity of the drag crisis. If the baseball suffered such a drag crisis at velocities common in baseball, anomalous effects could be important.[2] For example, a ball hit against the wind might go farther than a ball hit with the wind. It seems, however, that the baseball, usually rotating in flight, presenting different configurations of smoothness and roughness to the air as a consequence of the changing orientation of the stitches, does not correspond to any *uniformly* rough ball and does not undergo the sudden transition from smooth to turbulent flow that characterizes a drag crisis.

Since the drag for a rough ball can be less than for a smoother ball, even if there is no baseball drag crisis, the distance a batted ball will travel might still depend on the character of the ball's surface. A very rough, scarred ball with a surface that could induce

[2]Cliff Frohlich discusses the drag on baseballs—and much else about baseball—in an article in the *American Journal of Physics* published in 1984.

turbulence at low velocities could well travel farther than a new smooth ball. Smaller changes could also be significant. A change in the surface of the ball was initiated in 1974, when the traditional horsehide cover was exchanged for cowhide. A judicious estimate, however, suggests that for the range of surface conditions tolerated for baseballs used in the major leagues the dependence of the drag on the character of the basic skin surface can be neglected, i.e., cows and horses are not too different. But any significant changes in the height of the stitches might change the velocity at which turbulence begins and thus affect the flight of the ball and the distance balls can be hit.

Returning to the regulation baseball, the solid curve of Figure 2.2 shows the variation of the retarding force on the ball with velocity derived from the baseball drag coefficients of Figure 2.1. The force is expressed in terms of the weight of the ball; hence, the value 1.0 of the ordinate corresponds to the force of gravity. There may be significant differences in the drag force for different

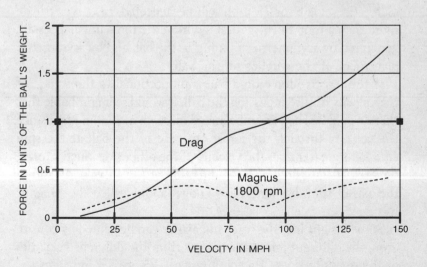

FIGURE 2.2   *The solid line shows the variation with velocity of the drag force on a baseball. The broken line shows the variation with velocity of the Magnus force for a ball spinning at a rate of 1800 rpm. The forces are expressed in units of the ball's weight.*

orientations of the stitch configurations with respect to the direction of motion, so the forces shown here must be considered as a kind of average over stitch orientations.

Note that the retarding force on the ball increases with velocity and is equal to the weight of the ball for a ball velocity of 95 mph. Therefore, the terminal velocity of a ball dropped from a great height is but 95 mph.

SPIN AND THE MAGNUS COEFFICIENT

The total force on a baseball, from the normal air pressure of 14.6 pounds per square inch, pushing the ball toward third base as it travels from pitcher to batter, is nearly 100 pounds. Of course, there is, ordinarily, a nearly identical force pushing the ball toward first base. If these forces differ by as much as an ounce and one-half—or about one part in a thousand—a ball thrown to the plate at a velocity of 75 mph will be deflected, or curve, a little more than a foot. Such modest, asymmetric force imbalances are generated by asymmetric spinning of the ball and by asymmetric placement of the stitches on the ball.

If the resistive force on a ball is proportional to the square of the velocity of the air passing the ball, it would seem probable that there would be such an unbalanced force on a spinning ball since the velocity through the air of one side of the ball at the spin equator is greater than the velocity of the other side. Such a force, directed at right angles to the direction of the air velocity and to the axis of spin, has long been known and is called the *Magnus* force.

Some insight into the force imbalance can be gained by considering the different forces that follow from the different velocities of the opposite sides of a spinning ball. Figure 2.3 indicates the origin of this transverse force for the "normal" situation, where the air resistance forces increase as the velocity increases. Consider that a curve ball is thrown (from left to right in the figure)

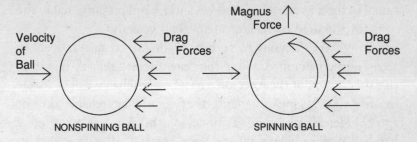

FIGURE 2.3 *The character of the drag forces on a nonspinning and spinning ball suggests the origin of the Magnus force. The imbalance in the components of forces normal to the surface of the spinning ball generates a force directed upward on the ball.*

by a right-handed pitcher at a speed of 70 mph so that it rotates 17 times counterclockwise (as seen from above) in its trip of about 60 feet from the pitcher's hand to the plate; such a ball will be rotating at a rate of about 1800 rpm; i.e., about one-half the rate of a typical small synchronous electric motor. The side of the ball toward third base (at the bottom of the figure) then travels about 15 times 9″ (the circumference of the ball), equal to 11′, farther than 60′, while the side toward first base travels 11′ less than 60′. The velocity of the third-base side is then about 82 mph, and the velocity of the first-base side is only 58 mph. As shown in Figure 2.2, the drag force increases with velocity; i.e., the difference in the pressure of the air on the front face of the moving ball is greater than on the rear face, and that pressure difference increases with velocity. We can then expect the air pressure on the third-base side of the ball, which is traveling faster through the air, to be greater than the pressure on the first-base side, which is traveling more slowly, and the ball will be deflected toward first base.

If the resistive drag force varies only as the square of the velocity and if the Magnus force is only an imbalance in that resistance, which follows from the faster motion of one side of the ball through the air than the other, we should expect that the Magnus force would be proportional to spin frequency, proportional to the air velocity (or ball velocity),[3] and proportional to the value of the drag coefficient at the ball velocity[c].

Though the Magnus force can be described qualitatively in terms of such an imbalance of the drag forces, a reliable quantitative description of the force on a stitched baseball is not available (more properly, measurements that should be reliable do not agree). Hence, we adopt a model of the Magnus force on a baseball that we believe must reflect, correctly, the general characteristics of the real force as expressed by the drag imbalance—and at worst is not likely to be seriously in error. So from this view the Magnus force, described by the imbalance of resistive forces on the ball that follows from the imbalance of velocity of the air flow past the spinning ball, is proportional to the rate of change of the resistance with velocity which is just the slope of the drag resistance curve shown as the solid line in Figure 2.2. From this model[d] of the Magnus force, the magnitude of the force (expressed in units of the weight of the ball) is shown in Figure 2.2 as a broken line for balls rotating at a rate of 1800 rpm, which is near the maximum for thrown balls. We judge the uncertainties in our estimates of the magnitude of the Magnus force on the baseball plotted in Figure 2.2 as about 25 percent.

The maximum Magnus force on a ball spinning at a rate of 1800 rpm is seen to be about one-third of the weight of the ball, so we cannot expect a ball spinning at that rate to curve more than one-third of the distance it will fall under gravity. Since the variation of velocity is proportional to the rate of spin, the Magnus force on a ball will be proportional to that spin rate, e.g., the force on a ball spinning at a rate of 900 rpm will be one-half that shown by the broken line in the figure.

Note that the Magnus force on a ball spinning at a constant rate increases with velocity up to speeds near 60 mph—the speed of a curve ball—and then probably falls off. According to this

---

[3]This relation was found in measurements by R. Watts and R. Ferrar and reported in the *American Journal of Physics* in 1987.

model of the force, for a given spin rate the Magnus force is smaller for a ball thrown at a velocity of 90 mph—such as a major league fast ball—than for a 60-mph ball. We might expect then that a hard-thrown ball will curve much less than a slower pitch, though the spin rates might be the same. The Magnus force can be expected to be less on the fast ball, and the force has less time in which to act as the ball reaches the plate quicker.

The forces that cause the ball to be deflected must also generate a torque that slows down the spin[e]. For a hard-hit ball traveling with an initial velocity of 110 mph, we use the model of the Magnus force we have adopted to estimate that the spin rate will decrease at a rate of about 20 percent per second. For a 400-foot home run, the backspin applied by the bat (perhaps 2000 rpm) would be reduced to about 650 rpm when the ball lands about 5 seconds later. This is consistent with experience, fly balls do not spin out of the glove of an outfielder when he barely catches the ball in the tip of his glove—sometimes called a "snow cone" catch.

DISTANCE OF FLIGHT OF A BATTED BALL

Projected by arm or bat, the distance a ball will travel can be calculated using the simple ballistic relations governing a body in flight and taking the retarding drag force in the direction of motion and the Magnus force, normal to the direction of motion, from the values shown in Figure 2.2. Typical trajectories are shown in Figure 2.4, and a graph of the maximum distance versus the initial ball velocity is shown in Figure 2.5. From these calculations, the maximum distance is obtained with balls projected at an initial angle of about 35° from the horizontal, though balls projected at 30° or 40° travel almost as far. Note that the ball falls at a rather large angle at the end of its flight; the trajectories are not symmetric. We find that the 385-foot fly ball hit at the

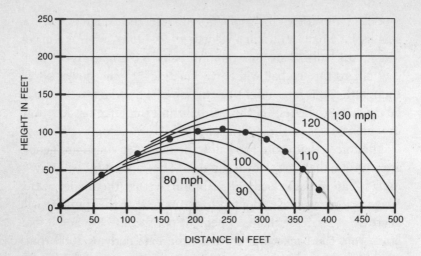

FIGURE 2.4 *The trajectories of balls projected at an angle of 35° with different velocities. The balls are assumed to be rotating with an initial backspin about one revolution per five feet—or 1800 rpm for a ball traveling 100 mph. The solid circles show positions of the ball at intervals of one-half of a second.*

optimum angle of about 35° will be in the air for about 5 seconds; a really high fly ball—or pop fly—will stay in the air more than 6 seconds. To put this into perspective, an average right-handed batter will run from home to first in about 4.3 seconds.[4]

Obviously these trajectories, and maximum distances, depend on the drag and the drag coefficients, which are imperfectly known. An uncertainty of 10 percent in the drag coefficient at

[4]A very fast left-handed batter will reach first base in about 3.7 seconds after a drag bunt, a few tenths of a second—a step—faster than an equally fast right-handed batter, while an unusually slow runner, such as catcher Ernie Lombardi of the 1940s, might take more than 5 seconds to go from home to first. In a timed racelike effort in 1921, Maurice Archdeacon circled the bases in 13.4 seconds; the legendary James "Cool Papa" Bell of the Homestead Grays is said to have made it in less than 13 seconds. To put these times into context, a world-class sprinter like Carl Lewis, running with track shoes on a carefully prepared track, would take about 11 seconds to run the 120 yards if the bases were laid out in a line and about 11.6 seconds to run in a straight line the 127 yards that a reasonable course about the bases would measure. Also see Chapter 4.

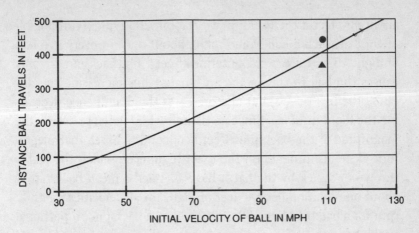

FIGURE 2.5 *The distance a baseball travels as a function of the initial velocity of the ball when projected at an angle of 35°. The solid circle and triangle show the distances as affected by a 10-mph wind blowing in and out.*

high velocities translates into an uncertainty in the maximum distances of flight of about 3.5 percent or an uncertainty of 14 feet for a 400-foot home run. Since we are primarily interested in *differences* in distance that follow from changes in conditions, such an accuracy is acceptable.

As a ball is thrown or batted for distance in a real situation, it usually has appreciable backspin (of the order of 2000 revolutions per minute), which generates a significant force perpendicular to the direction of motion and generally upward. Though the spin will probably increase the drag on the ball slightly, the Magnus force will somewhat increase the distance traveled by balls hit or thrown at smaller angles and reduce the distance of balls hit or thrown at larger angles. In general, the backspin can be expected to produce a modest increase in the distance the ball will travel as well as a decrease in the optimum angle of projection of a few degrees. According to my calculations, a ball launched without backspin at an angle of 35° and traveling 380 feet would travel about 400 feet if the backspin were about 2000 rpm. (Coming off of the bat, such a ball would rotate about once in every five feet of flight.) There are two countering effects: The backspin not only causes the ball to stay in the air longer, thus increasing the distance the ball travels; but also increases the drag, hence decreasing the flight distance. Since neither effect is known precisely, the magnitude of the backspin effect is uncertain. Since that magnitude depends in detail on the exact manner in which the ball is thrown or struck by the bat, I have assumed a mean backspin of about one rotation per five feet of flight for all velocities (or 1800 rpm for a ball traveling 100 mph); hence, a backspin proportional to the ball velocity is used in calculating the representative trajectories and distances given in Figures 2.4 and 2.5.

Obviously, the distance a ball will travel is strongly affected by the wind. Figure 2.5 also shows the effect of winds of 10 miles per hour (near the average wind velocity throughout the United States) on the flight of a 400-foot home run. With the wind behind the batter, the ball will go about 30 feet farther; with the wind against the batter, the 400-foot home run to center field will end up as a 370-foot fly ball out. In general, ball parks are laid out so that the line from "home base through the pitcher's plate to second base shall run East-Northeast"[5] so the batter will not face Nolan Ryan's fast ball in the late afternoon with the sun in his eyes. Hence, the prevailing westerlies of the Northern Hemisphere tend to blow out toward right-center field, helping the batter. Of course, for most ball parks, the effect of the wind is reduced because of the protection the stands afford.

Since the retarding force on a ball is proportional to the density of the air, a baseball will travel farther in ball parks at high altitude. A 400-foot drive by Cecil Fielder at Yankee Stadium, which is near sea level, on a windless summer day would

---

[5]*Official Baseball Rules*, Section 1.04.

translate to a 407-foot drive at Atlanta on the Georgia Piedmont, at 1050 feet the highest park in the majors before 1993. The same home run could be expected to go about 5 feet farther in Kansas City and 4 feet farther at County Stadium in Milwaukee or Wrigley Field in Chicago. These differences are not so great as to modify the game, but Fielder could expect his long drive to travel about 430 feet at mile-high Denver. And if the Major Leagues are further internationalized some day with a team in Mexico City, at 7800 feet, his blow could be expected to sail nearly 450 feet. Figure 2.6 shows the trajectories of long home runs hit with an initial velocity of 109 mph at an angle of 35° at sea level, at Denver, and at Mexico City. Old home run records will be swept away unless the fences are moved out in the high parks.

But even if the fences are adjusted, the high-altitude stadiums will still be a batter's boon and a pitcher's bane: with fences moved back, there will be acres of ground for balls to fall in for base hits, and, though the pitcher's fast ball will be about 6 inches quicker in Denver, the curve will bite about 25 percent less, which is more important.

With the smaller drag, the ball will also get to the outfielders faster in Denver than at Fenway Park in Boston. Indeed, a hard-

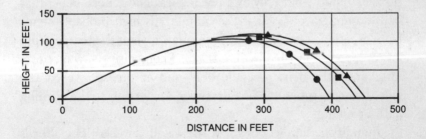

FIGURE 2.6 *Trajectories of home runs hit identically hard at an angle of 35° at sea level, at Denver, and at Mexico City. The position of the ball after 3, 4, and 5 seconds is also shown.*

hit "gapper," hit between the outfielders, will reach the 300-foot mark about 0.3 second sooner in Denver than at sea level, thus cutting down the range of the pursuing outfielder by 8 or 9 feet, a not inconsiderable amount in this game of inches. Even the range of the shortstop covering a line drive or one-hopper will be cut by nearly a foot in Denver.

The use of a less lively, "high-altitude" ball would reduce the altitude effect even as special, less lively, high-altitude balls are used in tennis, though for somewhat different reasons.

The air resistance also slows the bat down slightly; the swing that would drive the ball 400 feet at sea level would hit the ball about 404 feet if the bat met no air resistance and about 401 feet if the air resistance were reduced to that at Denver.

Temperature, barometric pressure, and humidity also affect the flight of a ball. The canonical 400-foot home run will go about 6 feet farther for a one-inch reduction in the barometer and as much as 20 feet farther on a hot 95° July day in Milwaukee than on a cold 45° April day[f]. The effect of temperature differences on the elasticity of the ball will also have an effect on the distance a batted ball travels.

Humidity per se has little effect on the flight of the ball. Indeed, since water vapor is a little lighter than air, if all other factors are the same, a ball will travel *farther*, if the air is exceptionally humid—but only by a few inches. The general belief that balls do not travel as far if the humidity is high probably stems from experience on windless humid nights when the temperature has dropped from the daytime highs. Then, with the cooler evening air a little denser—or deader—and no breeze to carry the ball, the home run in the hot afternoon carries only to the warning path at night.

Humidity certainly does affect the weight and elasticity of balls in storage, however. Balls stored under conditions of high humidity will gain some weight through the absorption of water from the air and their elasticity (the coefficient of restitution, discussed in Chapter 5) will be reduced[g].

The retarding force is proportional to the cross-sectional area of the ball, hence small balls will go farther than large balls. According to the *Official Baseball Rules,* a ball may be as large as $9\frac{1}{4}$ inches in circumference and as small as 9 inches, as light as 5 ounces and as heavy as $5\frac{1}{4}$ ounces. The stroke that propels the larger ball 400 feet will drive the smaller ball—of the same weight and elasticity—perhaps 6 feet farther. The stroke that drives the lighter ball 400 feet will drive the heavier ball—of the same size and elasticity—the same distance, give or take a foot or two. Though the heavier ball will come off the bat a little slower, its greater sectional density will carry it better through the drag of the air and just about compensate for the deficit in initial velocity.

TECHNICAL NOTES

a. The value of the Reynolds Number R for a sphere of diameter r moving with a velocity V through a fluid of density ρ and viscosity η, is:

$$R = \frac{\rho V r}{\eta}.$$

For a baseball, $R \approx 2200\,V$, where the velocity V is measured in mph.

b. The relation for the drag force F_d can be written as.

$$F_d = \frac{1}{2}C_d \rho A V^2$$

Here $A = \pi r^2$, with r the radius of the sphere, is the cross-sectional area of the sphere, ρ is the density of the air, V is the velocity of the ball, and C_d is the drag coefficient. For $C_d = 1$, this is just the force required to move the column of air displaced by the motion of the ball to match the velocity of the ball.

c. For velocities such that the drag coefficient C_d does not vary strongly with the velocity V of the ball through the air, the Magnus force, F_m, can be expressed as:

$$F_m = KfVC_d$$

where the force F_m is measured in pounds-force, the ball velocity V is expressed in mph, and the spin frequency f is measured in rpm. The results of Watts and Ferrar suggest a value for K such that $K = 2 \times 10^{-6}$.

Almost all of fluid dynamics follows from a differential equation called the Navier-Stokes equation. But this general equation has not, in practice, led to solutions of real problems of any complexity. In this sense, the curve of a baseball is not understood; the Navier-Stokes equation applied to a baseball has not been solved. Prof. Robert Romer, editor of the *American Journal of Physics,* told me of an eminent physicist who said, "There are two unsolved problems which interest me deeply. The first is the unified field theory [which explains the basic structure and formation of the universe]; the second is why does a baseball curve? I believe that, in my lifetime, we may solve the first, but I despair of the second."

Therefore, the above equation which I use to describe the Magnus force follows from the fundamental Navier-Stokes equation only after that equation is simplified through some rather drastic approximations. Hence, my simple Newtonian description of the complex processes that govern the curve ball does not contain all of the truth. But that description is *useful* and, surely, reasonably accurate for baseball velocities under 75 mph. In the absence of good measurements at higher velocities, we must be less certain of my estimate of the Magnus forces for balls traveling at greater speeds.

Incidentally, the Magnus effect responsible for the curve of the curve ball and the hop of the fast ball is not quite the same as the Bernoulli effect; it is more than the Bernoulli effect, which is why it is called the Magnus effect and not the Bernoulli effect.

d. To consider the Magnus force at velocities greater than 75 mph, we must consider the variation of the drag coefficient. In general, that coefficient, C_d, varies strongly with velocity near the transition between smooth and turbulent flow and we might reasonably expect the Magnus force (F_m) to vary as:

$$F_m = KfVC_d[1 + 0.5 \times (V/C_d) \times (dC_d/dV)]$$

This differential form follows naturally from the basic equations for both the low-velocity smooth flow and the high-velocity turbulent flow. The interpolative use of the relation for the intermediate region is not soundly based theoretically, but it accounts for the reversed Magnus force seen for smooth balls quite well. According to this relation, the Magnus force can reverse sign when the logarithmic differential $(V/C_d)(dC_d/dV)$ is less than -2 and the drag force then *decreases* as the velocity increases. Indeed, Briggs (and others) has reported anti-Magnus effects for balls with uniform surfaces at velocities which seem to correspond to smooth-turbulent transition velocities. The reduction in the Magnus force for baseballs at velocities near 100 mph that follows from these arguments (as shown in Figure 2.2) is the same effect that reverses the sign of the force for balls with more uniform surfaces.

e. Even as the interaction of the spinning ball with the airstream causes the ball to be deflected, there must be a reaction on the ball reducing the spin rate. We assume that the reactive torque L, which resists the spin, is:

$$L = kF_m r$$

where F_m is the Magnus force, r is the radius of the ball, and k is a proportionality constant that is not well known but which we estimate to be $\frac{1}{10}$. The energy-per-second (P) lost to the spin kinetic energy T is then:

$$P = L\omega = kF_m r\omega$$

where $\omega = 2\pi f$ is the angular spin velocity of the ball. The spin-down time constant τ is then:

$$\tau = T/P \text{ where } T = \frac{1}{5}mr^2\omega^2$$

where m is the mass of the ball. The time constant τ is a measure of the time required for the ball to lose its spin; for example, if $\tau = 5$ seconds, the spin rate will decrease by a factor very near $1/\tau = \frac{1}{5}$ in one second.

At high spin rates, the drag force on a baseball may increase. Some wind tunnel measurements show substantial increases in drag for spinning balls, some do not. In science—unlike many other matters—if two results differ widely, the truth is more likely to be found at one of the extremes than in the middle. I find that the game of baseball itself suggests that the drag on a baseball increases with spin, but not strongly. For moderate spins, where $v'/v < \frac{1}{2}$, I assume the drag is increased by a factor of $1 + (v'/v)^2$ where $v' = \omega r$ is the rotational velocity of the surface of the ball and v is the ball velocity. This form fits some measurements and has a theoretical basis. For larger values of v'/v I use a more arbitrary recipe, but that only affects pop flies and foul balls where there are also other uncertainties.

f. The drag resistance is proportional to the density of the air and hence, for a given temperature, to the barometric pressure; a reduction in pressure of $\frac{1}{3}''$ of mercury on the barometer decreases the drag by about 1 percent. On the average, the pressure—and the drag—decreases by about 1 percent per 275-foot increase in altitude. The density of the air is also reduced by about 1 percent for each 5° F rise in temperature, but the mean drag decreases a little more than 1 percent since the viscosity of the air will increase about $\frac{1}{2}$ percent (and the Reynolds number will decrease by about $\frac{1}{2}$ percent). For the canonical 400-foot home run, a decrease in the drag by 1 percent will add a little less than 2 feet to the ball's carry.

As a consequence of the reduced air density of altitude, the drive projected at an angle of 35° upwards at sea level that lands 400 feet from home plate will travel an extra 6 feet for every 1000 feet of altitude where conditions are otherwise the same. A 350-foot drive will gain an extra 4.6 feet per thousand feet of altitude. The linearity implied by the simple algorithm is good to a few percent; the overall accuracy, to better than 10 percent.

g. R. C. Larsen found that the weight of balls stored at 100 percent humidity for four weeks increased by 11 percent and the coefficient of restitution at an impact velocity of 25 mph decreased by 10 percent: when dropped on concrete from a height of 20 feet, the humidified balls will bounce only about 80 percent as high as the balls stored at low humidity. If that proportional decrease in elasticity holds at larger impact velocities, the swing of the bat that would drive a "dry" ball 380 feet will propel the ball stored at high humidity only 350 feet. After sitting in dry air for a few hours to dry the cover, the humidified balls cannot be distinguished from normal balls by a layman. However, an experienced pitcher would probably notice that they were heavier and softer than the balls he is used to.

PITCHING

THE PITCHER'S TASK

The most important person on the team, in any one game, is the pitcher.[1] The pitcher's job is simply defined—if not simply executed. From the pitcher's mound, 60′6″ from the rear point of home plate, he must throw the ball over the plate for strikes. But within that constraint he must project the ball in patterns of trajectories, velocities, and placements so that the batter cannot hit the ball squarely. But what is a strike?

A STRIKE is a legal pitch when so called by the umpire, which (b) is not struck at, if any part of the ball pass through any part of the strike zone.

The STRIKE ZONE is that space over home plate which is between the batter's armpits and the top of his knees when he assumes his natural stance [*Official Baseball Rules*, 1987, page 22].

1.05 Home base . . . [is] 17 inches long . . . with the 17-inch edge facing the pitcher's plate [*Official Baseball Rules*, page 5].

[1]Though over the whole season, the best hitter in baseball—Babe Ruth—was judged to be more valuable than one of the best pitchers in baseball—Babe Ruth. In 1919, Ruth was moved from the Boston Red Sox pitcher's box, where he worked every fourth day, to an everyday job in right field.

A typical six-foot-tall batter stands naturally at the plate with his armpits about 46″ above the plate. The tops of his knees will be about 22″ above the plate. With these definitions, a ball he does not swing at will be called a strike if it passes over the plate so that it knicks a five-sided box, with a cross-section the shape of the plate, 17″ wide, 24″ high, and 17″ deep, which might be suspended with the bottom edge 22″ above the plate. As a matter of practice, umpires define the high boundary of the strike zone about 4″ lower than the armpits, so the effective zone is about 17″ by 20″. Hence, the center of the ball must strike a three-dimensional target about 20″ wide by 23″ high and 17″ deep. Control pitchers hit corners with an uncertainty of about 3″. One must be a fairly good shot to shoot a pistol with that accuracy. A mistake, usually an error of a foot that puts the ball in the center of the strike zone, on one pitch of a hundred, may lose the ball in the bleachers—and the game.

The pitcher must not only throw the ball so that it passes through the strike zone, he must throw so that the batter does not hit the ball squarely. He does this by throwing the ball so that it passes through specific places at or near the strike zone that the batter finds difficult to reach, by throwing the ball at different velocities to upset the batter's timing, and by applying spin to the ball in such a manner that the ball passes the batter with different trajectories, confusing and confuting him. Thus, the pitcher varies the placement, the velocity, and the movement of the ball. The pitcher's action up to the release of the ball is part of the art of pitching; the action of the ball after release, determined by the laws of nature, is addressed by physics and is subject to our analyses.

THE CURVE BALL

Interest in the left-right curvature of balls sailing through the air is probably as old as ball games themselves. Isaac Newton, at the age of 23, discussed the curvature of tennis balls in terms that make good sense today. In the nineteenth-century genesis of mathematical physics, Lord Rayleigh analyzed the curvature of the path of spinning balls, and P. G. Tait, the eminent Scottish physicist, wrote extensively on the curves of golf balls—perhaps in an attempt to understand and cure a slice. The Baseball Hall of Fame in Cooperstown, New York, gives precedence to Candy Cummings as the first baseball pitcher to throw a curve ball, though the rules of the 1860s required Cummings to throw underhand rather as a softball pitcher today. Freddy Goldsmith and others confounded batters with curve balls at about the same time.

Balls curve as a consequence of asymmetries in the resistance of the air through which they pass. If the air resistance is greater on the third-base side than on the first-base side of a ball thrown from the pitcher to the batter, the ball will be forced—curve— toward first. Aside from generating curved paths, this resistance affects the flight of the pitched ball by reducing the velocity of a ball thrown from the pitcher's mound at a speed of 98 mph to a velocity of 90 mph as it crosses the plate about 0.40 seconds after it leaves the pitcher's hand. If the ball is not spinning very fast, during the time of flight it will fall almost three feet below the original flight line. If the ball is rotating quickly, differences in the force of the air on the ball transverse to the spin axis are induced that cause the trajectory of the ball to deviate from the original horizontal direction of motion and deviate vertically from the trajectory determined solely by gravity. Such asymmetric forces follow from the rotation of the ball when a curve or hopping (rising) fast ball is thrown or from differential forces on the seams for a slowly rotating knuckle ball. The action of a curve ball and knuckle ball are quite different.

First we consider the curve ball.

As we have noted, the force on a baseball is greater on the side of the ball that passes more quickly through the air due to the Magnus effect. The diagram at the top of Figure 3.1 shows the

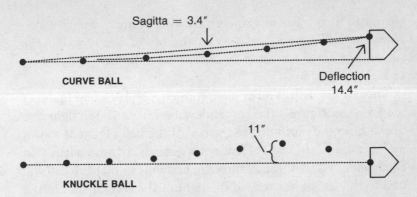

FIGURE 3.1   *The trajectories of a curve ball and a knuckle ball on their way from a right-handed pitcher to a batter. The curve ball is rotating counterclockwise as viewed from above the ball's line of flight.*

trajectory (reconstructed from Briggs's wind tunnel measurements) of a rather wide-breaking curve ball thrown so as to rotate counterclockwise—as seen from above in the figure—by a right-handed pitcher. This ball is thrown with an initial velocity of 70 mph, spinning at a rate of 1600 rpm, to cross the plate about 0.6 seconds later at a speed of about 61 mph. Although the radius of curvature is nearly constant throughout the ball's flight, the deflection from the original direction increases approximately quadratically with distance, i.e., four times the deflection at twice the distance. Halfway from pitcher to the plate, the ball has moved about 3.4 inches from the original line of flight, which is directed toward the inside corner and is moving toward the center of the plate. At the plate, the deflection is 14.4 inches and the ball passes over the outside corner. From the perspective of the batter—or pitcher—the ball that started toward the inside corner has "curved" 14.4 inches to pass over the outside corner. Moreover, one-half of the deflection occurred during the last 15 feet of the path to the plate. (We realize that the most useful curve from a tactical view curves down much more than sideways, but we discuss the transverse motion for expositional simplicity.) Does a curve ball then travel in a smooth arc like the arc of a circle? Yes. Does the ball "break" as it nears the plate? Yes. Neither the smooth arc nor the break is an illusion but a different description of the same reality.

Though the deflection as seen, correctly, by batter and pitcher is 14.4", the sagitta—the largest deviation from the straight line drawn from the beginning to the end of the ball's flight as shown in the figure—is but 3.4". Hence, it is difficult to throw a ball with a diameter of 2.9" through three aligned rods so that the ball will pass to the left of one rod, to the right of the second, and to the left of the third. In the course of arguments in 1870 as to whether a curve ball really curves, Freddy Goldsmith performed that feat in New Haven.

But Goldsmith probably threw a slower curve. For slow curves thrown with a definite spin rate, the deflection is approximately proportional to the time the ball is in the air. Hence, a ball thrown with an initial velocity of 65 mph with a 1600 rpm spin, which takes about 8 percent longer to reach the plate than the 70 mph pitch, will curve about 8 percent more.

Conversely, according to the model of the Magnus force we have adopted, it is almost impossible to throw a fast ball that curves strongly. As shown in Figure 2.2, the transverse Magnus force that induces the curved trajectory is smaller for velocities greater than 70 mph. Also, since the faster ball reaches the plate sooner, the force has a shorter time in which to act; for the same transverse force, the ball that travels 10 percent faster will curve 20 percent less.

There are other pitches of interest. The slider—sometimes called a "nickel curve" 70 years ago—is a kind of fast curve. Thrown at a higher velocity than the standard curve ball, the break of the slider is smaller than the deflection of the curve ball and the spin axis is such that the deflection is more nearly left-right than the curve—which, at best, is more of a pure drop. The screwball thrown by Carl Hubbell and others, called the "fade-away" by Christy Mathewson and a "scroogie" more recently by Fernando Valenzuela, is a kind of reverse curve thrown by a right-handed pitcher to break away from a left-handed batter a little like a left-handed curve ball.

Figure 3.2 shows typical spin directions for different pitches.[2]

THE KNUCKLE BALL

The thrown ball can also be deflected by the turbulence induced by the stitching on the flow of air passing by the ball. If the ball is thrown with very little rotation, asymmetric stitch configurations can be generated that lead to large imbalances of forces and extraordinary excursions in trajectory. Low-resistance turbulent flow can be induced by stitches on one side of the ball while the air will flow smoothly—and with larger resistance—past a smooth face on the other side. We can be more specific. Noting that the drag on a ball is proportional to the drag coefficient, we see from Figure 2.1 that the drag on a smooth ball thrown at a velocity of 65 mph is far greater than the drag on a rough ball. We might then expect that a ball that is smooth on one side and rough on

[2]It is interesting to observe the curvature in the flight of a table-tennis ball produced by throwing the ball so that it spins in the directions shown in Figure 3.2. Weighing one-sixtieth as much as a baseball, very small Magnus forces will generate substantial curves in the trajectories of the light ball over a span of 20 feet.

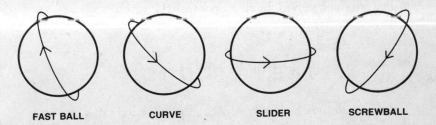

| FAST BALL | CURVE | SLIDER | SCREWBALL |

FIGURE 3.2 *Ball rotation directions, as seen by the batter, for pitches thrown almost straight overhand by a right-handed pitcher. The arrow shows the direction of rotation, which is also the direction of the Magnus force.*

the other will encounter an asymmetric drag force—larger on the smooth side—that will tend to deflect the ball toward the rougher, stitched side.

Measurements[3] of these asymmetric forces have been made on balls mounted in wind tunnels that allow the calculations of trajectories of a pitched ball. The diagram at the bottom of Figure 3.1 shows such a trajectory (albeit an exceptionally dramatic case). Thrown originally toward the center of the plate, when the ball is only 20 feet from the plate it is 11 inches off-center and heading toward the visiting team's dugout. The catcher starts moving desperately to his right to avoid a wild pitch, and the right-handed batter relaxes, knowing the ball will be two feet outside. Then, confounding catcher, batter, and pitcher, the ball ducks over the center of the plate for a called strike—and passed ball.

Other examples of a fluttering ball—which curves in both directions!—were also generated by the wind tunnel simulation. In general, the knuckle-ball pitcher tries to throw the ball with a small rotation (about one-half rotation from pitcher to batter) so that the stitch configurations—and forces on the ball—change on the way to the plate. Such a slowly rotating ball can be thrown off the knuckles—with the ball held between the forefinger and little finger—but the knuckle ball in practice is usually thrown off the fingertips or fingernails—which are carefully cut square.

The disadvantage of the knuckle ball from the view of pitcher, catcher, and manager is that the forces can vary strongly with very small differences in ball orientation. Hence, the pitcher, however skilled, finds it very difficult to control the pitch.

THE HOP OF THE FAST BALL

The fast ball may be the key baseball pitch; surely it is the most dramatic. How fast can a pitcher throw a ball? It seems that the

---

[3]These measurements were described by R. G. Watts and E. Sawyer in the *American Journal of Physics* in 1975.

fastest pitchers can throw the ball so that it crosses the plate with a velocity of about 100 mph. In 1946, Bob Feller threw a ball that was measured to have a velocity of 98.6 mph as it crossed the plate. In 1914, a pitch of Walter Johnson was found to have a velocity of 99.7 mph. Nolan Ryan threw a ball with a velocity timed at 100.7 mph. Pitches by J. R. Richard, Goose Gossage, and Lee Smith have also been determined to cross the plate with velocities near 100 mph. Among the old-timers, Lefty Grove and Smokey Joe Wood probably threw with comparable velocities according to contemporary judgments, and Satchel Paige and Slim Jones threw bullets in the Negro Leagues of the 1930s. Grove—and some batters—thought that he was faster than Feller, and Walter Johnson once said that "no one ever threw harder than Smokey Joe Wood" (Dr. Joe Wood after receiving an honorary degree from Yale in the 1970s). Though adequate velocity measurements are possible with instruments as simple as the ballistic pendulum used in 1914 and as sophisticated as the radar guns used today, some care is required to get an accurate result with either technique.

Since the ball slows down considerably on the way from the pitcher to the plate, the "muzzle velocity" of the ball—as it leaves the pitcher's hand—is about 8 mph greater than its speed across the plate. The ball loses speed at the rate of about 1 mph every 7 feet.

It is well known that the backspin applied to the overhand fast ball causes the ball to rise, or hop; such a ball will be thrown with a backspin of perhaps 1600 rpm and rotate about 10 times on its way from pitcher to plate. Though the hop is not likely to be much greater than 5 inches, this is more than enough to trouble the batter swinging a 2.5-inch diameter bat who must initiate his swing when the ball is about halfway to the plate and the deviation from the hop is only about 1 inch.

Like its cousin, the curve ball, the hopping fast ball follows a smooth arc on its way to the plate, but half of the hop deviation occurs in the last 15 feet of that flight.

Figure 3.3 shows the trajectories of a typical 90-mph fast ball

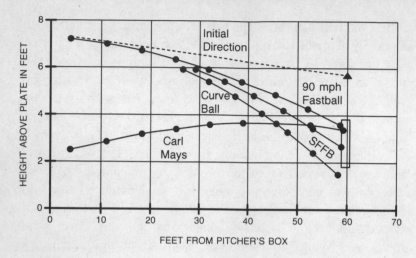

FIGURE 3.3 *Elevation views of the trajectories of pitches from the pitcher to the batter. Ball velocities are measured in mph, rotations in rpm. The points show the positions of the ball at intervals of one-twentieth of a second. The different horizontal and vertical scales distort the trajectories.*

thrown "high and tight" together with trajectories of a curve ball at the knees and a split-finger fast ball (SFFB), all thrown from the pitcher's hand in the direction marked by the straight line. The trajectory of an underhand pitch, such as the blazing fast ball Carl Mays threw from his shoe tops in the 1920s, is also shown to demonstrate the variety of slants the batter must face. Figure 3.4 shows the trajectories, undistorted by the scale differences, of certain pitches as they cross the plate.

The relatively small hop of about 5 inches may seem somewhat surprising. But the "upward curve" of 5 inches is not very different from the "tailing off" seen in fast balls thrown sidearm (or from a three-quarter overhand pitch), in which the axis of rotation is tilted so the hop has a left-right component that is more clearly seen. Again, the small break follows from the short time the force acts and from the smaller Magnus coefficient at higher velocities—as shown in Figure 2.2. A 75-mph (high school) fast ball, thrown with the same motion—although with less intensity—

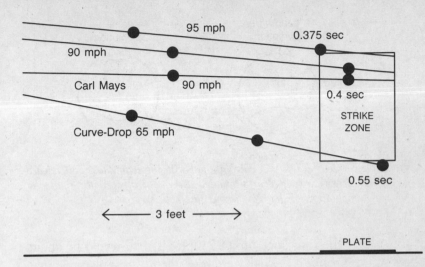

FIGURE 3.4 *Trajectories—with no scale distortion—of different pitches as the balls cross the plate. The 90-mph and 95-mph fast balls were launched with the same initial trajectory. The ball images show positions at times that differ by one-fortieth of a second.*

may actually hop more since the Magnus force (as shown in Figure 2.2) may be larger and the force would be applied for a longer time to the slower ball.

The magnitude of the Magnus forces and drag forces presented in Figure 2.2 represents a kind of average over different stitch configurations. Since the pitcher can control those configurations, he can produce somewhat different effects by releasing the ball so that it spins with the axis oriented differently with respect to the stitch pattern. The diagram at the left side of Figure 3.5 shows a ball held for the delivery of a "with-the-seams" fast ball, in which two seams pass the equator at every revolution. At the right, the release position for the "cross-seams" fast ball is shown and four seams pass the equator per revolution.

It is plausible that the aerodynamics of the ball delivered with such different spin axes will be significantly different. An inspection of the graph of Figure 2.1 illustrating the variation of the drag coefficients with velocity for balls of different smoothness

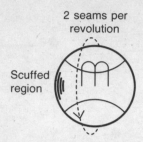

FIGURE 3.5 *At the left is a conventional with-the-seams fast ball release configuration with a scuffed area situated asymmetrically. At the right, the release configuration of the cross-seams fast ball is shown.*

shows that the resistance on a 95-mph fast ball would be about two-thirds greater if the ball were very smooth. It is plausible that the with-the-seams rotation, showing only two seams per revolution, has a smoother-than-average profile in the airstream and that the cross-seams rotation might appear rougher than average. If this is the case, we could expect a greater-than-average drag on the with-the-seams fast ball and, hence, a somewhat smaller-than-average drag on the cross-seams delivery.

How large can we expect such effects to be? Using the "average" drag resistance we estimate that a pitcher must deliver the ball with a muzzle velocity near 98 mph if the ball is to cross the plate with a velocity of 90 mph (an average speed of 94 mph). If the drag were reduced by half for the cross-seams delivery, the ball would cross the plate with a velocity of 94 mph (and an average velocity of 96 mph) and reach the batter almost a foot and a half faster. If the drag were increased by two-thirds, as for a *perfectly smooth ball,* the ball thrown with an initial velocity of 98 mph would be slowed down to only 82 mph at the plate (for an average velocity of 90 mph) and reach the plate about 2 feet behind the "standard" pitch. But these are extremes, holding for an ideally smooth ball and a particular uniformly rough ball. Real baseballs are far from either. It is plausible, however, that differences in cross-the-plate velocities as great as 2 mph might occur because of the different spin axes that result from different ball grips. That would translate to a different average velocity of about one mph, or a little over 6 inches on the fast ball.

Since the curve, or hop, on a ball follows from a difference of the drag on the two sides of the ball, we might expect that the spin axis that generated less drag would also generate less curve, or hop. Then, if the cross-seams fast ball moves faster, it might hop less.

Some pitchers purposely throw a "sinking" fast ball, with little spin and that about an axis that is relatively ineffective in generating a Magnus force. Such a pitcher might hold the ball as shown in the left-hand drawing of Figure 3.5, except with his fingers set on the smooth leather rather than the stitches so the ball will slide from his hand with little spin. This ball, delivered without much backspin, will undergo little upward Magnus force and come to the plate as much as 4 to 6 inches lower than the hopping fast ball. Delivered at the knees, the batter will tend to hit the top of the ball and roll out weakly to the infielders.

There are long-standing arguments about the magnitude of the rise of an overhand fast ball. Can such a pitch, thrown with a lot of backspin, actually rise as it heads toward the plate? From the trajectories shown in Figure 3.4, the answer would seem to be no. But, as we have pointed out, we do not understand the ball's flight through the air as well as we would like. Could our arguments be wrong, and could the players who swear that some pitchers throw a rising ball be right?

We can cast some light on the problem by noting that since the 90-mph fast ball must fall almost 3 feet on its way to the plate if there is no hop, the effect of the backspin must be such as to "curve" the ball upward about 3 feet if the ball is to rise.[4] But the magnitude of that hop will be duplicated by a pitcher who throws a 90-mph fast ball sidearm, with a sidespin instead of a backspin, only that curve will be from left to right (for a right-

---

[4]Indeed, since over-the-top pitchers must throw the ball slightly down to deliver a strike at the belt, the ball must curve up as much as 3 feet more than the 3-foot gravity effect if it is to actually rise as it crosses the plate in the strike zone.

handed pitcher). But who has seen a fast ball curve 3 feet? In fact, a hard fast ball thrown sidearm (or with a three-quarter motion) will tail off no more than 6 inches, if that much. Therefore, overhand fast balls certainly do not rise, and our calculations are not likely to be wildly wrong.

Good softball pitchers, throwing underhand upward to a batter 46 feet away from a pitcher's box that is not raised on a 10-inch mound, *do* throw balls that rise as they pass the batter's armpits. However, the underhand baseball pitchers such as Carl Mays in the 1920s and Dan Quisenberry and Kent Tekulve in the 1980s, who throw 60'6" from an elevation of 10 inches, can't really get the ball to rise as it passes the batter.

If the baseball fast ball falls nearly 3 feet on its way from pitcher to plate, how can thoughtful, intelligent players possibly believe it might rise? Here we have a matter of perception. If one draws a straight line from the pitcher's hand to the position of the ball as it crosses the plate as shown in Figure 3.3, the peak of the ball's actual trajectory will lie only about 8 inches above that line. Over the distance of about 56 feet from hand to plate, the real path then differs only about 8 inches from a straight line. The 3-foot drop is from the line projected from the ball's flight as it leaves the pitcher's hand. (This is the line that would describe the ball's flight if the game were played in outer space, where there is no gravity or air.) The player who sees the ordinary fast ball as traveling almost in a straight line is not wrong. We can see that he might reasonably consider that a ball that arrives above the usual line by 6 inches must rise.

Part of the reason for the misunderstanding among players about the "rising fast ball" stems from the foreshortened view of the ball trajectory from the pitcher's mound, the catcher's box, and the batter's box. If the player were to watch the fast ball from the on-deck circle, he would see it fall a lot (about 1 foot in 8) as it crosses the batter at the waist. The falling trajectory of the fast ball can also be seen occasionally on TV when the cameraman chooses to show the ball crossing the plate as seen by a camera at right angles to the line from pitcher to plate.

Similar misunderstandings occur about the flight of a throw from third to first by a strong-armed third baseman. Some players state emphatically that a hard-thrown ball travels in a straight line, but a physicist concludes that the ball must drop about 10 feet. Again, the 10 feet is the drop below the initial line of projection. The peak trajectory of the ball is only about $2\frac{1}{2}$ feet above a straight line drawn from the third baseman's hand to the first baseman's mitt. And a $2\frac{1}{2}$-foot deviation over the 127-foot distance between third and first is not far from a straight line. Moreover, since fields are landscaped with the center of the diamond higher than the base paths (so the infield will drain toward the foul lines) and the pitcher's mound is 10 inches high, the ball thrown and caught at a height of 6 feet can pass near the ear of the six-foot-tall pitcher. And the pitcher may well have to duck to avoid the catcher's throw to the shoe tops of the second baseman in position to tag the runner attempting to steal second.

The ordinary overhand fast ball is thrown so that the ball rolls off the first and middle fingers of the pitching hand—indeed, these fingers usually rest on the stitches to accentuate the grip and the spin. If the pitcher places the ball so that he holds its axis between his two widely split fingers at about the first joint, he will propel the ball almost as fast but with very little spin. The curves of Figures 3.3 and 3.4 show trajectories of such well-thrown split-finger fast balls.[5] Usually the split-finger fast ball is not projected quite so efficiently as the regular backspinning fast ball and does not achieve quite the same velocity. With the same initial trajectory as the hopping 90-mph fast ball directed across the letters, the 80-mph split-finger fast ball reaches the plate about 16 inches lower and at the knees—and about 7 feet behind the fast ball.[6]

---

[5] If the ball is jammed further toward the hand, the velocity of the pitched ball is reduced considerably and the pitch is more of a pure change-of-pace. Such a pitch is often called a "fork ball." However, in practice, the term is used variously.

[6] If they were released by two pitchers at the same time, the split-finger fast ball would lag behind the fast ball by 7 feet as they crossed the plate.

PITCHING WITH AN ILLEGALLY MODIFIED BALL

From baseball's beginnings, pitchers have enhanced their skills at throwing the ball with different velocities and spins by modifying the ball and the pitching hand. Saliva, sweat from the forehead, or surreptitious Vaseline applied to the pitching hand aid in the delivery of the slowly rotating "spitball" by allowing the ball to slip from the hand without spin. The main effect of the lubricant is that it allows less skilled pitchers to emulate what a skilled knuckle ball pitcher can accomplish legally—though the spitball can be thrown harder than the knuckle ball. The application of a lubricant to the forefinger and middle finger on delivery of an overhand fast ball can produce some of the same effects as the split-finger fast ball: The ball will slide off the fingers, accumulate less backspin, and then drop more than the usual fast ball thrown with backspin.

Scarring or scuffing the ball can produce asymmetric forces on the ball that result in aberrant trajectories. Since it is probably impractical to scuff or scrape the surface of the ball so that the imperfection has as dominant an effect as the stitching, it might seem that such modifications cannot be very important, and that is largely the case for casual throws. But the highly skilled pitcher[7] can throw the ball so that the effect of the stitches is symmetric, but if the ball is scuffed on one axis—and not the other—unbalanced forces can be realized, which act only in one direction, as suggested in the left-hand diagram of Figure 3.5. Properly thrown, the scuffed left-hand side of the ball could induce low-resistance turbulence in the air passing by while the air will pass the smooth right-hand side in high-resistance smooth flow and the ball will veer to the left—*toward* the scuffed area! Since a devia-

---

[7]This scuff-ball effect can also be simulated by a table-tennis ball thrown indoors. To conduct this test, take a square of Scotch tape—perhaps $\frac{3}{8}'' \times \frac{3}{8}''$—and stick it on the ball so it wrinkles. Then throw the ball with backspin as suggested by the left-hand diagram of Figure 3.5 with the tape on one, or the other, pole of the axis of rotation. The ball will veer, counterintuitively, *toward* the tape[a].

tion of a fraction of an inch can change a home run into a pop fly or a double-play ground ball, the controlled deviations a skilled pitcher can induce by disfiguring the ball are important.

The curve—called "swing"—of the bowled cricket ball plays a part in cricket like the curve in baseball. Throwing on the run, the bowler, projecting the ball with an overhand straight-arm motion, sends it to the batter on the bounce—nearly as fast as a fast ball in baseball. The cricket ball is about the same size and weight as a baseball, but the raised stitches form a circle about the ball's equator. Also, the bowler is allowed to shine one side of the ball, leaving the other side rough. He then relies on a kind of scuff-ball effect to project a ball that swings, evades the batter, and knocks down the wicket—though knuckle-ball–like effects, where the seam catches the air, are also important.

PITCHING IN THE WIND—AND ON HIGH

Excepting the recent intrusion of domed stadiums, baseball is an outdoor game and affected by the vagaries of the weather, which influences the mechanics of pitching and the facility of pitchers. Here we concern ourselves only with the weather's effect on the flight of the pitched ball, leaving the benefits of hot weather on pitching arms to physiologists and psychologists. Moreover, in the contest between pitcher and batter, we consider only the effects of weather on the pitched ball, leaving the problem of wind-blown fly balls and balls in the sun to outfielders and managers.

The weather affects the flight of the pitched ball mainly through the velocity of the wind over the diamond. Though small effects follow from variations of temperature and barometric pressure—the fast ball is quicker by an inch or two on a hot day and at the lower air pressure before a storm or at a higher-altitude ball park—these will not affect the game significantly. The average wind velocity over the United States is surprisingly constant at about 10 mph. (The wind surely blows harder at certain places, such as Mount Washington, but major league baseball games have not been regularly scheduled there.) Moreover, the prevailing wind is from the west and will usually be in the pitcher's face. Of course the weather is famously perverse everywhere—especially at Candlestick Park in San Francisco, where a balk was called on Stu Miller in the 1961 All-Star Game when the wind blew him off the mound—and winds can come, betimes, from any direction.

The wind's effect on the pitched ball is subtle and important to pitcher and batter. It is again convenient to consider specifically the effects of a 10-mph wind, though on the playing field of most major league parks the protection of the stands usually reduces the average wind velocity to a somewhat smaller level. The effects we discuss all vary almost linearly with wind velocity, hence, a light 5-mph breeze will have half the effect of the "standard" 10-mph wind.

As an example of the consequences of wind, we consider a standard major league fast ball, thrown overhand, that crosses home plate at a velocity of 90 mph 0.40 seconds after leaving the pitcher's hand with a muzzle velocity of 97 mph. We also assume the pitcher applied a typical backspin to the ball of 1800 rpm, which causes it to curve up and reach the plate perhaps 5 inches above where it would have without the spin. With the 10-mph wind in the pitcher's face, the ball will reach the plate slightly later (lagging by about 3 inches of flight path), traveling a bit slower at about 89.3 mph, and will arrive at about the same height. With the wind at the pitcher's back, the ball will reach the plate about 3 inches sooner at a velocity of 90.7 mph and cross the plate about an inch lower than it would without the wind. The ball will actually hop slightly less. These effects are all small compared with the precision of pitcher or batter, and not very important. But a crosswind can be a little less innocuous. If the wind is blowing across the field at a velocity of 10 mph, the fast ball will be blown about 3 inches—the diameter of the ball—off course at the plate, enough to trouble the nibblers who are trying to hit corners if the wind is erratic, as it often is. Again, the effects are roughly proportional to the wind velocity.

If the fast-ball pitchers can largely ignore moderate breezes, curve-ball and junk pitchers have a lot more trouble. We consider specifically a left-right deflection of a wide-breaking curve thrown so that it crosses the plate with a velocity of 61.2 mph in calm weather 0.6 seconds after leaving the pitcher's hand with a velocity of 70 mph. This particular ball, thrown with a sidespin of 1800 rpm, breaks about 15 inches. Thrown initially toward the inside of the plate, the ball will break over the outside corner. (Actually, the major league pitcher is usually more interested in the vertical deflection—down—than the transverse break.) With the wind directly behind him, or in his face, there is little significant change. The ball reaches the plate about 9 inches sooner (in flight distance) with the wind behind him and about 9 inches later with the wind in his face, but that is not important. The ball also breaks about the same amount, though here our lack of knowledge of balls near this speed leaves some uncertainty. A very slow curve may well break more when it is thrown against the wind.

A crosswind, however, modifies the pitch considerably. If the wind is from the north, from third to first, the right-handed pitcher's curve will break about 21 inches toward first base, but if the wind is from the south, the break will only be about 8 inches. Hence a gusty day with crosswinds can cause serious problems with slow pitches. If the wind kicks up, a pitch thrown accurately at the corner of the plate can end up a ball or cross right over the center of the plate to be deposited in the bleachers by the batter as a souvenir for a fan.

The pitchers will also be hurt at high altitude. A mile high at Denver, the fast ball will take a little less time to cross the plate—and gain about 6 inches—but the curve will break about 25 percent less. A curve that will break left-right about 8 inches and drop an extra 8 inches (due to the overspin component) at sea level will break about $1\frac{5}{8}$ inches less and drop about 4 inches less in Denver. The ball breaks less because it crosses the plate faster and thus has a little less time *to* break; in addition, the

Magnus force is smaller. Similarly, the knuckle ball will dance perhaps 25 percent less.

While moving the fences back may reduce home runs at Denver, the large outfield that would result, combined with the reduced break on the pitches, leaves Denver as a pitcher's purgatory, if not quite hell.

THE ENERGETICS OF A THROWN BALL

A player's actions in throwing a ball hard are complex, and the analysis of the throw lies more in the realm of physiology than physics, but the results of a calculation of the projected ball's total energy and an estimate of the rate of energy transferred to the ball, i.e., the power, in throwing provide some illumination of the character of the process.

In this estimate, we consider a pitcher throwing a major league fast ball so that it leaves his hand with an initial velocity of 97 mph to cross the plate about 0.4 seconds later with a velocity of 90 mph. We estimate the distance between the point where the ball begins its motion toward the plate and the point of release as 8 feet; that is, the pitcher holds the ball about 4 feet behind the pitcher's rubber as he begins to bring the ball forward toward his point of release, which is about 4 feet in front of the rubber. The released ball then has an energy of about $\frac{1}{6}$ horsepower-seconds. (A horsepower-second is equal to the energy output of a one-horsepower motor running for one second—and that is enough to lift 550 pounds one foot.) Making the crude—but useful—approximation that the force the pitcher applies to the ball is constant, we find that the throw takes about 0.11 seconds[8] and that the average force on the ball of about 12 pounds generates a mean acceleration of the ball of about 40 g's—forty times

---

[8]This is not the "delivery" time. The time between commitment—after which a balk would be called if the pitch were interrupted—and release is more like 0.8 seconds.

the acceleration of gravity. With these numbers, we find that the average power the pitcher transmits to the ball in the course of the pitch is about 1.5 horsepower! Since the pitcher's body is also put into motion by the contraction of his muscles (in fact, his hand and wrist are moving nearly as fast as the ball when it is released), we conclude that his musculature must have generated energy at a rate exceeding 3 horsepower during the action of throwing. About 20 pounds of muscle are required to generate one horsepower, hence such power can only be generated by the large muscles of the thighs and thorax. As pitchers well know, it is important to develop and retain leg and body strength to throw the fast ball.

Although the flow of energy from the pitcher's body to the ball is surely complex, and the description of that flow is not within the scope of this book, some comments about a particular aspect of the energy transfer are in order. In the course of the throw, the elbow leads the hand and ball at the midpoint of the action. At this time energy is stored in the stretching of the arm's tendons. That energy is transferred to the ball in the last portion of the throw as the spring-energy of the stretched tendons is released. When age or injury reduces the tendons' elasticity, the arm goes "dead."

In general, the elasticity of all tissues decreases with age, and the blazing fast ball of the young player must be replaced by the craft of the older pitcher. Late in his long and remarkable career,[9] Warren Spahn, commenting on his life in baseball, said, "When I was young and threw the fast ball. . . ." At 45, when he retired from baseball, Spahn was probably about as strong as he was at 25, but his arm was not as elastic.

---

[9]Spahn won over 20 or more games in a season 13 times, the last in a 23 and 7 season in 1963 when he was 42.

TECHNICAL NOTE

a. The nominal similarity between the action of the table-tennis ball and the scuffed baseball may be a little misleading. Since the Reynolds number for the small ball, traveling at rather low velocities, is only about 10,000 (corresponding to a baseball velocity of only 10 mph), the separation of the boundary layer, important in the baseball knuckle ball and scuffed ball, is probably overshadowed for the table-tennis ball by the classical Bernoulli effect, that is, the air flowing over the Scotch tape has a longer path and must move faster, thus generating a smaller pressure that results in a deflection of the ball toward the tape. The similar effect of the speeding up of the air passing over the protuberant baseball stitches would not seem to generate sufficient force to account for the deflections of the heavier baseball.

RUNNING, FIELDING, AND THROWING

RUNNING

Baseball players run. They run the bases, and they run in the infield and outfield in fielding balls. To understand this split-second sport we must have some understanding of how fast they run.

We can begin to understand how fast-ball players run by reviewing how fast sprinters on tracks run—their speed is surely a limit for baseball players. Figure 4.1 shows record times for sprints as of 1987 where the straight line fits a simple model that can be simply described by noting that a world-class track athlete like Carl Lewis, running with track shoes on prepared tracks, can sprint at a rate of about 36 feet per second after a start out of starting blocks that takes 1.0 second.[1] An equally fast man, given a running start from 30 yards behind Lewis—so that he crossed the starting line as the starter's gun went off—would be about 1 second, or 11 yards, ahead of Lewis at the 30-yard mark.

---

[1] The extrapolation of world record times given by the straight line in the graph predicts a time for Lewis, on a track, of 4.3 seconds for a 40-yard dash. Hence, many of the 40-yard dash times reported from football training camps must be taken with a substantial dose of salt.

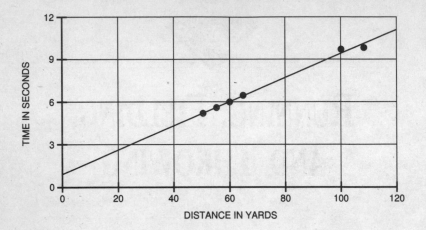

FIGURE 4.1: *World-record times for sprint distances in 1987.*

We take the speed of a very fast outfielder, running on grass in a baseball uniform with baseball shoes and carrying a glove, as 30 feet per second. But how much time does he lose in starting? Unlike Lewis, the outfielder generally doesn't even know in what direction to run at the moment the bat hits the ball. He can probably tell whether to go left or right after the ball goes about 40 feet, which takes about 0.3 seconds. If he is stationed 300 feet from home plate, he will first hear the crack of the bat about that time—about 0.27 seconds after the bat-ball collision. Not only does he lack starting blocks, but he is not at that time even leaning in the direction he has to go—he doesn't yet know where that is. I would have to conclude that even Cool Papa Bell, perhaps the fastest man to ever play the outfield—who could probably run almost as fast as Carl Lewis—would lose 1.5 seconds on his start in the outfield.

On the bases, without a glove and free to swing their arms, the speedier players will run a little faster—say 32 feet per second, and they will be able to start a little quicker. But if the base runner is not to be caught leaning by the pitcher's pick-off throw, even the quickest will still not start as fast as Lewis. We estimate that Lou Brock still lost 1.2 seconds upon starting.

With these recipes, we can calculate the distance the fastest player can cover in a given time; those distances versus time are shown in Figure 4.2 along with distances Carl Lewis might attain. The exact values are not too reliable for short times but are probably accurate for times over 2 seconds.

JUDGING FLY BALLS

How do players judge fly balls? Just what did Tris Speaker, Richie Ashburn, Mays, and Mantle, as well as the kids of my youth (and Little Leaguers today), do to "judge" a fly ball? How do players know where to run to catch the ball that sails toward them? An android with a computer brain might calculate trajectories, but I suspect that humans follow simple learned patterns derived by experience; a human can beat supercomputers on pattern recognition. But to compute, or operate from a set of memories, the player—human or android—must work from information pro-

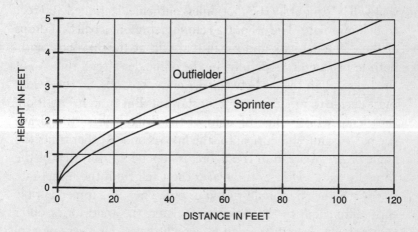

FIGURE 4.2: *The upper curve shows distance vs. time for a very fast baseball player in the outfield or on base, while the lower curve shows the prediction for a world-class sprinter in a race on a prepared track.*

vided by the senses. Hence, without knowing just how the information the fielder receives is processed by the brain, we can still discuss the limitations on that information and understand much about judging fly balls.

To discuss running down fly balls quantitatively, one should have some idea as to how long balls stay in the air and how much ground players can cover in that time. The trajectories of the balls and their flight times can be calculated rather accurately from the known air resistance of the balls; some fly-ball trajectories are shown in Figure 2.4. And with the aid of Figure 4.2, we can estimate how far outfielders can run during those flight times. But how does the player decide where to go? How can he determine where the ball will come down?

What information does the player need to estimate the distance the ball will go? We begin with the ball hit straight at the player—the most difficult hit to judge. Our first conclusion is that the initial rate of take-off of the ball as seen by the fielder doesn't help. Balls hit different distances look the same initially. Figure 4.3 shows the trajectory of three fly balls where the position of the ball is shown by markers every second; one represents a fly ball that will be caught by the motionless outfielder stationed 300 feet from home plate. The other trajectories represent a ball that drops at the 250-foot mark and a ball that falls at the 345-foot mark, both to be caught, hopefully, by the outfielder after a run. All of the balls stay in the air between 4.7 and 5 seconds. So the fielder must decide to run in, run out, or stay put. But from his position, the changes in the line of sight of the three balls as they leave the bat are almost identical for the first second. At that time, the angle of rise, marked in the figure, is about 15.5° for each of the three fly balls.[2] Hence, the player can't tell from the initial rate of rise whether he should run in or out. There are imperceptible early differences in the trajectories; later, the trajectories differ substantially. After two seconds—or a little less for a very percep-

---

[2]The three trajectories that are described here were taken from a continuum with the same rise in the first second.

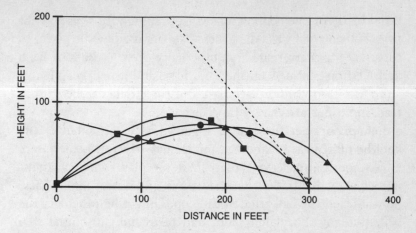

FIGURE 4.3: *The trajectories of three fly balls hit different distances directly at an outfielder will look the same to him for the first second of travel. The markers show the position of the ball at intervals of one second. The broken line shows the angle of observation of the descending ball just before the catch.*

tive player—the outfielder knows to run in or out from his assessment of the line-of-flight, which now is appreciably different for the different balls. Having hesitated about two seconds, however, he can't cover much ground; he won't get a good jump. Starting to run after two seconds have elapsed, from Figure 4.2, it is clear that he must run hard to cover 45 feet in the remaining three seconds to catch either the short or long ball.

Of course, the experienced outfielder uses other clues. He watches the swing—is it a wheelhouse power swing or an off-balance swing by a batter half-fooled by the pitcher? How did the hit sound? Was it the "crack" of the squarely hit ball or the "clunk" of the mis-hit? Perhaps he makes (unconsciously) an estimate of the distance of the ball from its perceived size. But this is probably no help during the first second. The differences in the effective size of the ball one second after it is hit on the trajectories shown in Figure 4.3 are smaller than the acuity limits imposed by diffraction in the iris of even a perfect eye. At these distances, depth perception follows from subtended sizes and not from binocular effects. All of these data are used by the player to construct a pattern that, from experience, he can use to make a useful estimate of how far the ball will go and hence, get a better jump to run forward or backward. But these data can be hard to read and a mistake is very expensive.

Let us consider the ball of Figure 4.3 that is hit over the fielder's head so as to land—if unimpeded—45 feet behind him, 4.65 seconds after leaving the bat. If the player guesses almost immediately that this is a long ball, turns, and starts running back, he will reach the ball after an easy lope. But if he misjudges the ball, thinking it will fall short, and takes but one initial step forward, he is in trouble. He will start forward and stop before starting back, losing an extra start and stop time of about 1.8 seconds. Moreover, he must not only retrace the extra 5 feet he moved forward, but also cover 50 feet to where the ball will land. We read off from the graph of Figure 4.2 that he won't quite reach the ball in the 2.85 seconds that is left, and the ball he should have caught will sail over his glove for a double.

The opposite misjudgment of the soft fly—in my grandfather's day, a "can of corn"—hit so as to land in front of the outfielder, about 250 feet from home plate after a flight of 4.7 seconds through the air, leads to a similar penalty. If the ball is judged correctly, it should be caught easily after a 50-foot jog. But, again, a too-precipitous misjudgment is disastrous. If the player starts back, and then must turn and run forward, he won't quite get to the ball, which will fall for a single.

For the well-played ball, during the time the ball is in the air, the estimate of the distance it will carry is continually refined so that the fielder is near the correct position when the ball is about one second from the ground, whence alignment procedures take over, directing player and glove to a precise conjunction with the ball. I doubt that the brain does much in the way of calculation, but I expect that it compares the new pattern with the store of patterns from the experience of past catches. Hence, the value of experience; hence, the practice.

During the last alignment second, the player tries to position himself so that the falling ball moves exactly at him, as shown in Figure 4.3. If the line of sight of the ball drifts upward (as for the ball over his head), the player moves back; if the line of sight falls (as for the ball that would land in front of him), the player trots forward.

Balls hit to the left or right are much easier to judge, and errors in estimated distance are punished much less. With the ball hit to the side, the fielder can, while he begins to run to the left or right, observe the trajectory of the ball much as one looks at the pictured trajectory of Figure 4.3—albeit somewhat foreshortened—and he can differentiate between the trajectories just as the reader does.

With the left-right question solved in the first half-second of flight, how does the fielder proceed? Whether it is teenagers shagging softball flies on a playground or major leaguers fielding fungos in spring training, we see that almost all balls that are not hit almost directly to the player are caught on the run. For a high, lazy fly hit to land not too far from the player's position, the run is an easy lope timed so the fielder and ball meet as the ball returns to about the height of the player's eye, five or six feet above the field. If the ball is a line drive with little hang time or a long drive to the warning path, the player may have to run very hard, but, again, he and the ball meet at the point of the catch. How does the player plot the course that results in that precise juncture?

Part of the answer is that the player runs laterally so as to keep the left-right motion of the ball directly at himself. If the ball is hit initially to the fielder's right and, in the course of the running catch, the ball drifts to the right, the fielder will speed up; if the ball drifts to the left, the fielder will slow down. The direction or angle of the run is more difficult to set but is designed such that the falling ball—somewhat after it reaches the apex of its flight—moves, again, almost exactly at the player. Later, when the ball is coming down, if the ball drifts up too fast, the running fielder will turn toward the fence; if the ball drifts downward, the fielder will turn toward the infield. Hence, for nearly a second before the catch, the ball is coming right at the moving fielder with no motion to the left or right and no motion up or down. Thus, only one pattern needs to be learned, one that is valid no matter how fast the player is running—if he is running at all.

We can gain some insight into the process by reviewing Figure 4.4, which shows an ideal route by a broken line, and a practical route by a solid line, for a running catch of a fly ball. In particular, we discuss fielding the long fly ball with the trajectory shown in Figure 4.3.

Generally, the fielder runs the path, shown as the solid line in Figure 4.4, at a speed such that he always sees the ball move directly at him as far as left-right motion is concerned. Hence, ideally, at each one-second interval he is looking at the ball in the direction shown by the dotted line of Figure 4.4; in the system

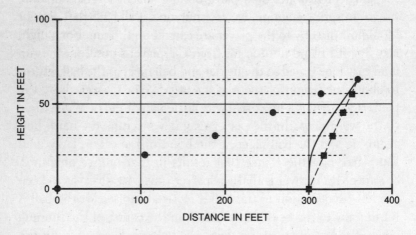

FIGURE 4.4:   *The circles show the position, at one-second intervals, of a long fly ball (as shown in Figure 4.3) hit 345 feet toward the power alley, and the position of a right fielder making a running catch where the player, whose initial position was 300 feet from home plate, has judged the distance of the ball perfectly. The lateral dotted lines show the line of sight of the player watching the ball at one-second intervals. The solid line shows the ideal trajectory of the running fielder: he starts directly to the right about one-half second after the bat hits the ball, when he knows that the ball is going to his right but not yet how far it will go.*

of the fielder the ball's left-right movement is null.

To catch this long ball, hit about 345 feet at an angle of 10° with respect to the line from the fielder to home plate in the direction of the right-field power alley, the right fielder must run about 75 feet in about 4.5 seconds; from Figure 4.3, we can see that that is well within the compass of an average outfielder. Note that the fielder may even slow down at the end of his run, since the ball is now traveling slower.

However, even for balls hit to the left or right, it's not easy to know how far the ball will go, especially in the first second of flight, and the fielder makes a mistake. Let us assume that he misjudges the ball seriously and actually starts in toward home plate, later correcting himself. Now he must run about 15 feet farther, and he will lose a few tenths of a second in making the sharp turn; a fast outfielder will probably still be able to make the catch.

In summary, the misjudgment that changes ordinary running catches into more difficult, but still makeable, catches for the ball hit 10° away from the fielder also changes easy catches into base hits for balls hit directly at the fielder. Stopping and turning around is much more expensive in time than running to the side at the wrong angle.

THROWING IN THE INFIELD AND OUTFIELD

We know that some pitchers throw harder than others: the variance in the velocity of the fast balls of different major league pitchers is more than 10 percent. A few of the best pitchers today—such as Gooden, Clemens, and Nolan Ryan (past 40!)— throw the ball so that it crosses the plate moving as fast as 95 mph. Conversely, a few very good pitchers, such as Vic Raschi with the Yankees in the 1930s and 1940s and John Tudor in the 1980s, threw fast balls that crossed the plate traveling no more than 85 mph.

Without the benefit of measurements, we can assume that position players also throw with similar velocities. Since a good "gun" is an absolute necessity for catchers, third basemen, and shortstops, we need only consider differences at these positions between good and very good arms. Figure 4.5 shows the elapsed time as a function of distance for balls thrown with an initial velocity of 100 mph—a throw from a player with a very good arm. Let us assume that the difference between the "guns" and just good arms is about 5 percent, where the fastest deliver the ball with an initial "muzzle" velocity of about 100 mph. With such a delivery, a ball thrown 135 feet by the shortstop or third baseman to the first baseman's glove will take about 1.09 seconds. A player who throws the ball with an initial velocity only 5 percent less will deliver the ball to the first baseman about 0.06 seconds slower. During this time the fastest base runners will have covered about 2 feet, but we know that 2 feet is significant.

Of course if a player with only a moderately good arm can release the ball 0.06 seconds more quickly, he can make up for the advantage of the big gun. He can't gain much on the actual

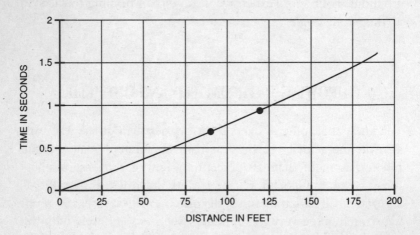

FIGURE 4.5: *Elapsed time vs. distance for a ball thrown with an initial velocity of 100 mph. The distances from home plate to third base and from home plate to second base are marked.*

throwing, which takes about 0.10 seconds for all players; however, a player generally takes about 0.9 seconds from the time the ball sticks in his glove to the beginning of his throw, so a quick player can pick up time there. This quick release is more important for second basemen, even as the difference in the time that the ball takes to go the shorter distance to first is less important. This is particularly important for the double play: Bill Mazeroski, considered by many the best ever among second basemen at turning the double play, augmented a good arm with about the quickest release seen in baseball. Extrapolating from world-record times in track, we find that Carl Lewis, running on a track with track shoes from a starting block, can run the 90 feet from first base to second in about 3.5 seconds—and 78 feet in 3.17 seconds. If a very fast prospective base-stealer on first can take a lead of 12 feet and start as fast as Lewis—but runs (in baseball shoes and in a baseball uniform) 10 percent slower—he will slide into second 3.5 seconds after a dead start—if he loses nothing on the slide. Every 3 feet of lead is worth about one-tenth of a second, and a rolling start is worth a good half-second. Indeed, the difference between the runner having his weight mainly on his front foot and mainly back foot (but don't let the pitcher catch you leaning!) must be worth more than one-tenth of a second. Hence, the cat-and-mouse play between pitcher and the runner on first.

From the delivery to the catcher's mitt takes about 0.45 seconds for a fast ball; perhaps 0.03 seconds longer for a slider or one extra foot for the base runner; and an extra 0.13 seconds for a curve and four feet for the runner. But the throwing motion—defined as the time after which the pitcher, throwing from the stretch, can't change his mind and escape a balk—is about 0.8 seconds. Hence, the total time from stretch to catcher is about 1.3 seconds. Some pitchers (Jack Morris is on the slow side) are as much as 0.2 seconds (or six running feet) slower. And a few relief pitchers seem to be significantly faster—perhaps as much as 5 running feet.

If the catcher throws with an initial velocity of 100 mph (which is about the absolute limit), the ball will travel the 128 feet to second base about 1.0 seconds after it is released. But, like the pitcher, the catcher needs time—typically about 0.9 seconds—to get the ball out of his mitt and complete the throwing motion. Adding the times together, 1.3 seconds for the pitcher throwing a fast ball to the catcher and 2 seconds from catcher to second base low and on the first-base side of the bag in place for the tag, we have 3.3 seconds for a perfect play and the base runner is out. But give a fast man a good start—worth about 0.2 seconds—and no one can throw him out. If the throw is only slightly off, the runner has an even better shot at stealing; for every foot the shortstop or second baseman has to move to make the tag, the runner gains another two feet or more.

Some say it is easier to steal third than second—but since the tactical gain is not so great, stealing third is uncommon. The throw to third is short and the ball gets there about 0.7 seconds after the catcher's release—about 0.3 seconds faster than to second. That means about a 10-foot difference for the runner, who can take a larger lead at second. If he can take a lead better by more than 10 feet than he could take at first, stealing third is easier than stealing second.

Since outfielders throw for longer distances, the differences in their throwing velocities lead to considerable differences in the time-of-flight, which in turn has significant tactical effects on the game. Moreover, hitting plays a somewhat larger role in the whole craft of the outfielder than for second basemen, shortstops, and catchers; hence, outfielders with relatively weak arms, like Ralph Kiner, are tolerated if they hit—and with Kiner, who hit 369 home runs in ten seasons, much more than tolerated.

Since outfielders throw on the run, they can probably throw with muzzle velocities that run as high as 10 mph greater than for infielders handling a normal chance and throwing from a set position. Since throwing plays a less central role than other skills, there is also probably a greater variance in the throwing velocities of outfielders than even for pitchers. We expect that Roberto Clemente, throwing on the run from center field, let the ball loose traveling about 110 mph; perhaps some of Kiner's throws from left field started out with velocities no greater than 90 mph. (Here we note that in a domed stadium with no wind, the player who can *only* throw the ball at 90 mph can throw a ball about 310 feet in the air while the rare player who can reach 110 mph can throw the ball from home plate to bounce off of the center-field wall 400 feet away.)

Although there are a variety of tactically important throws where velocity is important, we choose to examine just one special—and specially dramatic—throw: the throw that is designed to catch a runner at home plate who is tagging up after a catch. We also add that it is the last of the ninth with the score tied, so the outfielder will not have to hit a cutoff man. To be definite, we have the fielder catching the ball on the run, so that his throw starts from a point 300 feet from home plate. Moreover, we assume that he plans—wisely or not—to hit the catcher on the fly with his glove on the ground ready to make the tag.

Figure 4.6 shows the different trajectories that must be used by players throwing the ball with different initial velocities. The man who throws very, very hard—with a velocity of 110 mph—throws "on a line" with a low trajectory (and at an elevation angle of

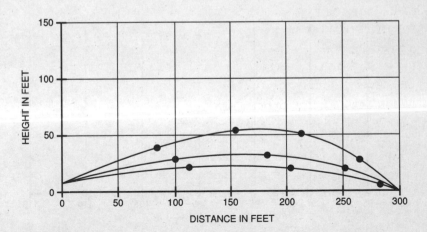

FIGURE 4.6: *Trajectories of throws with different initial velocities that travel 300 feet. The solid markers show the position of the ball every 0.4 seconds.*

about 13°); the man who has a relatively weak arm and lets go of the ball with a velocity of 90 mph must throw the ball with a substantial arc (and an initial angle of elevation of about 27.5°). He loses time not only because his ball is moving slower but because his ball must go nearly 10 percent farther.

The differences in time-of-flight between the hard and weak throws are substantial. From Figure 4.7, we see that the very strong throw gets to the plate in about 2.65 seconds. If the player takes 0.6 seconds to release the ball, it will get to the catcher about 3.25 seconds after the catch and the runner will be out. Few can go from third to home after a catch much faster than 4 seconds. However, the man with the good arm, throwing with an initial velocity of 100 mph, will have the ball in flight about 3.1 seconds and the ball will get to the plate about 3.7 seconds after the catch. Unless the throw is absolutely perfect, the catcher will probably not get the fast base runner, though he may be able to tag the slower player. But the ball from the 90-mph thrower will get to the plate no sooner than 4.4 seconds after the catch; hence, the game is lost. Of course, in all of this we assume that the ball to be fielded is a soft high fly so that the fielder can take a bead on

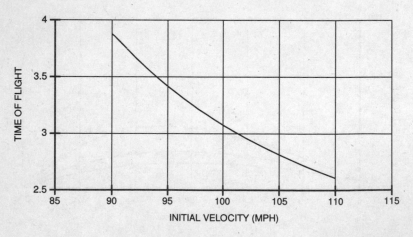

FIGURE 4.7: *Time-in-flight for balls thrown 300 feet with different initial velocities.*

it and catch it while running in toward the plate. If the catch were made while the fielder were stationary, for example, one would have to add as much as 0.5 seconds to account for a slower release and a slower ball.

Thus, 300 feet is too far; indeed, most teams station their outfielders no farther than 250 feet from home plate under such a circumstance.

We note that those players with weaker arms might be better off throwing at a lower angle to get the ball to the plate on the bounce. If the surface is Astroturf, the 90-mph player can gain as much as 0.2 seconds, or 6 feet, on the runner by throwing on the bounce. But if his team is playing on grass and his grounds keeper has kept the grass long and well watered to help his team (which relies on singles, speed, and baserunning), the ball may lose so much speed at the bounce that nothing will be gained.

BATTING THE BALL

A MODEL OF BATTING

There are many ways of batting the ball successfully through swinging at it so that precision of placement of the batted ball rather than extreme velocity is important. The drag bunt and the hit-and-run behind the runner at first base, for instance, represent purposeful hits that are made by swinging with less than full power. Hits to the opposite field often fall into this category. Conversely, the batter is often fooled by the pitcher so that his timing is thrown off and he swings weakly—albeit sometimes successfully hitting, perhaps, a "Texas leaguer" or an infield grounder that bounds so slowly that he beats the throw to first. In the following discussion, we do not consider the imperfect swings—perhaps a majority in the game—but only the full swing made with maximum effort. Only a few players—Babe Ruth and Ted Williams come to mind—*always* seemed to make full swings.

In considering the mechanics of the batting process, it is useful to construct a model of the full swing that is tractable but sufficiently close to a real swing by a real batter so that the consequences of the simple model illuminate the complex reality. Such modeling is simplified by the realization that a player swings a bat very much like a weight on the end of a rope; to a very good approximation, the forces on the bat exerted through the hands are directed along the axis of the bat. Hence, if the motion of the hands is known throughout the swing, the motion of the bat can be determined uniquely.

A model swing at a waist-high pitch by a strong player is illustrated in Figure 5.1, which shows a plausible trajectory of the bat throughout a swing in a horizontal plane calculated from simple dynamics and the trajectory of the batter's hands. The bat

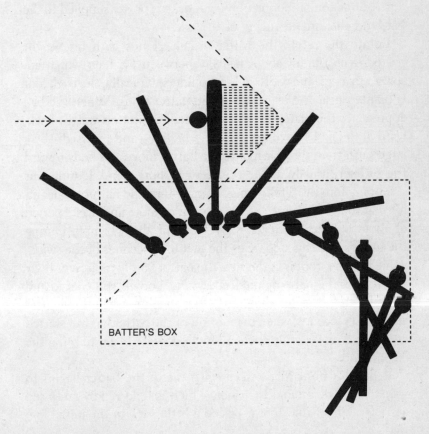

BATTER'S BOX

FIGURE 5.1 *The trajectory of the hands and bat during a typical swing. The position of the bat is shown at intervals of one-fiftieth of a second.*

velocity is taken as the velocity of the radius of gyration with respect to the end of the bat which is near the vibrational node or "sweet-spot" of the bat—the best place to hit the ball. In the figure—and calculation—the initial position of the bat lies in the plane of the swing for convenience though most batters actually hold the bat more nearly upright. That difference does not seriously affect the consequent bat trajectory.

The graphs in Figure 5.2 show the variation with time during the swing of the velocity of the hands, the velocity of the barrel of the bat, the total force on the bat and the component of force in the direction of motion of the hands, the power supplied to the bat, and the kinetic energy of the bat.

Before the pitch, the batter usually stands with his weight placed over a point about halfway between his feet, which are spread to about the width of his shoulders; typically, the back foot is planted near the rear edge of the batter's box. When the ball approaches the plate, the batter generally shifts his weight backward an inch or two. This motion is halted—and reversed—by a large push from the rear foot as the batter moves his body toward the pitcher, usually in the course of a short step.[1] During this "stepping into the pitch," the strong, 180-pound batter pushes off his rear foot with a force of 300 pounds or more and over a period of about 0.2 seconds reaches a velocity of about 6 mph. During the first part of this step—i.e., the first 0.15 seconds of the swing, while his front foot is in the air—the batter hardly rotates his body and rotates the bat only moderately. His body is still cocked and he is looking over his shoulder at the incoming ball. Only after about 0.05 seconds, when the motion of the batter has just started and he has moved only about an inch, does he begin to bring the bat around.

When his front foot is securely planted, the batter begins to rotate his body around that foot, which is held rigidly and used as a pivot. The front foot is placed a little back of the initial line

---

[1]Babe Ruth often took two steps, beginning the first step about the time the ball left the pitcher's hand.

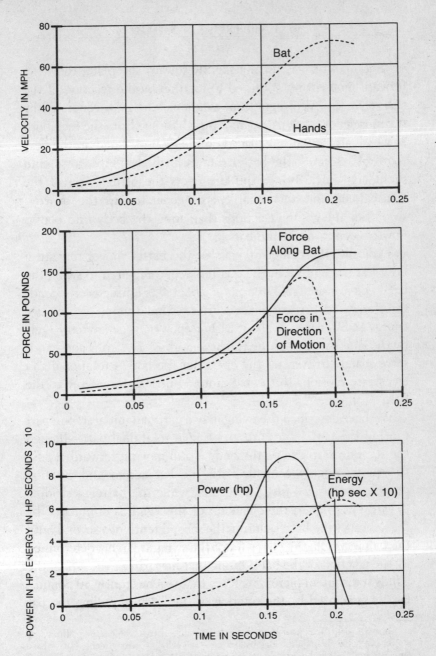

FIGURE 5.2   *The variation with time of the velocities (top), forces (center), and power to the bat and bat energy (bottom) during the model swing.*

of motion, which is usually slightly toward the plate: thus, the forward momentum is turned to rotation by the reaction of the stiff front leg. Also, significant rotational energy is generated by the muscles of the torso acting through torque from the feet, both of which are now firmly placed on the ground. Then, in the next tenth of a second, the batter rotates his body—hips, torso, and shoulders—as he brings the bat across the plate. Most of the translational and rotational energy generated in the quarter-second of the complex actions that bring the body into play is now concentrated in the bat.

Typically, the center of mass of the batter moves forward a little less than a foot in the quarter-second required to swing the bat. The considerable energy of about 0.6 horsepower-seconds transferred to the bat in a quarter-second (hence, an *average* energy transfer rate of about $2\frac{1}{2}$ horsepower) is generated largely by the large muscles of the thighs and torso. The arms and hands serve mainly to transfer the energy of the body's rotational and transverse motions to the bat and add little extra energy to the bat.[2]

We have described the swing in two parts, an initial step-in part moving the batter forward, and a final rotational part—though the rotation starts while the body is still moving forward. Those two parts roughly coincide with the bat action and reaction phases. During the first 0.1 second, while the batter is striding toward the pitcher, the character of the swing is dominated by the action of the batter; during the second tenth of a second until the ball is struck, the character of the swing is largely determined by the reaction of the bat. During that first part of the swing, the hands transmit an increasing force that reaches nearly 50 pounds, largely generated by the rotation of the body in pulling the bat

---

[2]In particular, the contribution of the hands and wrists to the energy of the bat is almost negligible. Though the long-ball hitters—generally big, strong men—usually have large, strong hands and wrists, there are exceptions. No National League player has ever hit more home runs in a season than Hack Wilson, 5'6" tall and 190 lbs. Built like a fire plug, Wilson wore size $5\frac{1}{2}$ shoes, and his small hands, "which matched his feet," could not have been especially strong—but the tremendous strength in his legs and torso was transferred efficiently to the ball through his strong arms.

about an arc. At the end of that time, the hands and the trailing bat reach a velocity of about 20 mph, and the kinetic energy of the bat is about $\frac{1}{20}$ horsepower-seconds. The weight shift during this time is equivalent to a motion of the body toward the pitcher with a velocity of about 6 mph corresponding to a kinetic energy of perhaps 0.6 horsepower-seconds for a 180-pound batter. This energy is stored in the motion of the body.

After 0.1 seconds elapse, the bat moves out of the arc of the hands and can be considered to exert an increasingly large reaction force on the hands and arms that will reach a value near 200 pounds as the bat crosses the plate. During this time, that large force straightens (uncocks) the wrists. The force from the reaction of the bat, transmitted through the hands and arms, slows down the motion of the body, which is now mainly rotary; in the sweet swings of the great batters, the stored kinetic energy of the body is transferred efficiently to the bat,[3] increasing the velocity of the bat from 20 mph to about 70 mph. The rate of that energy transfer exceeds 8 horsepower about 0.14 seconds after the initiation of the swing—and about 0.04 seconds before hitting the ball. This 8 horsepower is far greater than that actually generated by the contraction of muscles during this period. For many batters, the energy transfer from body to bat is nearly complete: the forward motion and rotation of the body is stopped almost completely when the bat crosses the plate. Of course, if the motion of the body were uniformly reduced by a factor of three-fourths as the kinetic energy of the body was transferred to the bat, the bat would still pick up 94 percent of that energy.

There has been some controversy over the relative importance in batting of rotational motion and translational motion. Sometimes this controversy has been connected to Ted Williams, who

---

[3]The part of the total force of reaction by the bat in the direction of motion of the hands is responsible for the energy transfer. That force ranges up to 150 pounds, requiring strong arms. This is why Pete Gray, the one-armed outfielder who played for the St. Louis Browns in 1944, hit no major league home runs. Using only one arm, he could not apply a force as great as the 150 pounds necessary to transfer enough of the energy developed in his strong thorax and legs to the bat so that he could hit the ball very far.

has emphasized the importance of rotation, and Charlie Lau, who emphasized translation. Of course, both are essential (as both Williams and Lau knew) and they are interrelated—even as the energy of translation of the body goes into the rotary energy of the bat. To those who consider translation unimportant, consider a batter standing with his back foot on a platform set on ball-bearing wheels on a toy train track running parallel to a line from the pitcher to the plate and with another such platform to catch his front foot, if he should be able to step forward. This batter would not be able to stride forward but he could still rotate. But how far could he hit a ball? I say not much past second base.

But if the tracks ran perpendicular to the pitcher-plate line, the batter would be able to stride forward but unable to rotate. So handicapped, without the possibility of a strong rotation, again, he would be lucky to clear second base. To hit a baseball with dispatch, one needs both to step into the ball and to rotate.

Typically, the fast ball struck by the bat carries about one-fifth as much kinetic energy as the bat. If the ball is struck squarely, about half the energy of the swinging bat is transferred to the ball in the impact, so the speed of the bat is sharply reduced, by about 30 percent. About one-third of the original bat-and-ball energy is carried off as kinetic energy in the flight of the ball from the bat, and the rest of the energy (about three-tenths of the original total energy) is lost in friction in the course of the distortion of the ball—and then to heating the ball.

The course of the swing after the ball is struck is kinematically irrelevant. The emphasis by coaches and sports teachers on the "follow-through" in baseball (and also in golf and tennis, for instance) is designed to ensure proper actions *before* the ball is struck. Figure 5.1, however, does show typical positions of the bat and the hands after the ball is struck; the pattern would be a little different if the ball were missed.

The graphs in Figure 5.2 show a typical variation of the bat and hand velocities during the swing, the variation of the forces applied to the bat, and the energy transfers during this model swing. Though the precise quantities shown in the graphs derive from our particular model and must be quite special, the qualitative values of the quantities and the character of the changes of the quantities that are presented are probably close to that of real swings.

The batter in the diagram is assumed to have swung in such a manner as to drive a ball to center field as far as he can. If this right-handed batter[4] should miscalculate the velocity of the pitch and hit the ball 0.005 seconds early, and to right field, he would hit the ball before maximum bat velocity was realized and generally not hit the ball quite as hard. Since the loss of energy after the maximum is reached is usually small, if he swings too quickly, by 0.005 seconds, he will lose little power in driving the ball to left field.

A pull hitter swings so as to maximize the bat velocity a little later in the swing. In general, the pull hitter has more time and distance to apply force to the bat and can therefore transfer a little more energy to the bat and hit the ball a little harder. If the right-handed batter is to make a minimal adjustment in the position of the arc of his swing, he will naturally hit a ball over the inside portion of the plate later in the swing and to left field. He will hit a ball on the outside of the plate earlier in the swing and to right field. It is then harder to hit a very fast ball over the inside of the plate, and "high and tight" is a natural strikeout pitch for an exceptionally fast pitcher.

Aiming for the center-field bleachers, the batter in Figure 5.1 will swing the bat at a speed (of its prime hitting region) of about

---

[4] Though the right leg bears more of a burden than the left for a batter who swings from the right side, the arms share the forces nearly equally. Hence, there is no natural advantage for a right-handed man in batting right-handed. Indeed, since the left-handed batter is favored in many ways, many players who throw right bat left: the batter who swings from the left side is closer to first base and moves in that direction as he finishes his swing, for some reason right-field fences tend to be closer than left-field fences, and there are more right-handed pitchers (whose slants are easier for a left-handed batter) than left-handed pitchers. There are, however, two famous first basemen who threw left and batted right, Hal Chase and George Bush. Chase, a great-fielding first baseman, was considered the best ever at the position by Walter Johnson and Babe Ruth, while Bush, captain of a fine Yale team that went to the finals of the NCAA tournament, did an exemplary job of throwing out the first ball on opening day.

70 mph so that the bat, passing through the hitting zone, is effectively rotating about a point near the handle end of the bat. If he swings as much as $\frac{1}{100}$ of a second early, the ball will go foul down the left-field line; if he is $\frac{1}{100}$ of a second late, the ball will go foul into the stands down the right-field line.

Batting is an intricate art, and there are various ways to bat well. If the batter in Figure 5.1 swings in such a manner (more of an "arm" swing) that his hands move faster through the hitting zone and the bat is effectively moving in a longer arc, perhaps as if it were rotating about a point about one-half a bat length beyond the end (toward the player), the bat will be rotating slower at the time of impact and the error of $\frac{1}{100}$ of a second, early or late, will result in a long fair ball to left or right field rather than a foul ball. With such a swing, less precision in timing is required, but the batter must begin the longer motion a little sooner than for the more compact arc of the "wrist" hitter.

Often, tall, strong hitters like Darryl Strawberry and Dave Winfield use a longer sweep and longer arc to balance their early swings (and better chance of misjudging the pitch) with the greater power generated in a longer swing and the less precise timing required in a sweeping swing. They also can afford to use a heavier bat. Smaller players, e.g., Howard Johnson, who hit with power more often tend to rely on quick, precisely timed swings with shorter arcs. They will choose a lighter bat to help them swing more quickly. With such swings, they can delay their commitment and better judge the ball.

Most distance hitters hold the bat at the end to maximize the bat velocity realized from the whip motion. Babe Ruth can be considered to hold the bat beyond the end! Batting left-handed, for much of his career he held the knob of the bat in the palm of his right hand.[5]

Players of yesteryear who emphasized precision, like Wee Wil-

---

[5]Ruth, like most players, often made minor changes in his batting style. Pictures of the Babe on his follow-through after hitting his sixtieth home run off Tom Zachary in 1927 clearly show the knob of the bat below his right hand.

lie Keeler, who succeeded by "hitting them where they ain't," often used long arcs and reduced the angular velocity. Ty Cobb and Honus Wagner separated their hands on the bat and swept the bat through the hitting zone, adding to the angular accuracy of their hitting (though Cobb, especially, would often bring his hands together early in the swing).

THE BALL-BAT COLLISION

For a well-hit ball, the bat-ball collision takes place near the *center of percussion* of the bat. An impact at this point on a free bat transfers no momentum—or force—to the handle. Conversely, a force at the handle is not transferred to the point of impact. Also, the collision with the ball takes place over a time of about $\frac{1}{1000}$ of a second, during which the contact point on the bat moves only about an inch and the handle moves a much smaller distance. The position of impact near the center of percussion, together with the small motion of the handle, means that at the moment the bat strikes the ball, the bat can be considered as a free piece of wood, with the clamping effects of the grip of negligible importance. It is this property of the swing of the bat that allows us to divide the action neatly into the complex physiological actions that transfer energy from the body to the bat and the relatively simple processes that transfer energy from the bat to the ball. We address only the bat-and-ball interaction here.

Very large forces, reaching values as high as 8,000 pounds, are required to change the motion of the $5\frac{1}{8}$-ounce ball from a speed of 90 mph toward the plate to a speed of 110 mph toward the center-field bleachers in the $\frac{1}{1000}$ of a second of bat-ball contact. Hence, for a long home run, the force on the ball reaches a value near 8,000 pounds with an equal reactive force on the bat. Such forces distort bat and ball: The ball is compressed to about one half of its original diameter; the bat is compressed about one fiftieth as much. The collision is not elastic; much of the energy of ball and bat is dissipated in frictional heat.

Treating the bat as a moving free object with known weight, weight distribution, and velocity, and knowing the weight and velocity of the ball, one can calculate the initial velocity of the batted ball from the laws of mechanics given the inelasticity of the ball and bat. The ball may be considered as a spring; the bat applies force to the ball, compressing it, and the ball exerts force on the bat on regaining its original contours. The recoil from this exerted force propels the ball away from the bat.

Figure 5.3 shows the results of measurements by Paul Kirkpatrick of the force versus compression distance for a simulated bat-ball collision and similar diagrams for the compression of a golf ball and the compression of the bat. (Complete *stress-strain* measurements were not made on bats, and the curve shown for bats is an estimate presented to show graphically the small energy loss by the compression of the bat.) The upper curves of each set plot the force versus distortion distance for compression; the

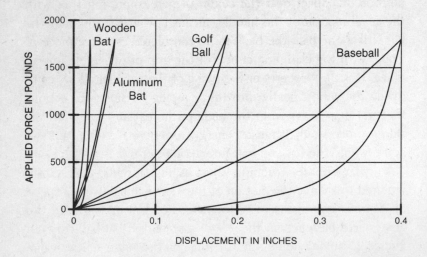

FIGURE 5.3 *The distortion of balls and bats as a function of the applied force for the cycle of compression and expansion. The displacement here is the change in the diameter of the bats or balls.*

lower curves show the force versus distortion distance for the following expansion. The area under the upper curve is proportional to the energy absorbed by the ball in motion; the area under the lower curve is the energy returned by the ball pressing against the bat, in resuming its spherical shape. For an ideal spring, the two curves will coincide; the energy released from the spring upon expansion is equal to the energy absorbed upon compression. The area enclosed by the two curves is proportional to the energy dissipated or lost in friction. For the ideal spring this is zero. As shown here, in expansion, the baseball returns only about 30 percent of the energy supplied in compression. The golf ball, much closer to a perfect spring, returns more than 75 percent of the compressive energy when struck by a driver.

The wooden bat is actually more elastic than the ball,[6] but since the wood is hard and the compression is small, the energy stored by the wooden bat is only about one-fiftieth of that stored by the ball. If the wood were just as elastic as the ball, the energy that distorted the bat would be returned as efficiently as the energy of distortion of the ball, and the hardness of the bat would be irrelevant. If the wood making up the bat were so weak that the distortion was not returned at all, i.e., the ball would put a permanent dent in the bat, all that energy of distortion would be lost and the ball would not leave the bat with quite as much velocity.

Years ago, players used to hone their bats with a hambone because they believed this would make the bat harder and make the ball fly off the bat faster. Sometimes, batters would (illegally) hammer nails into the hitting area so the ball would strike iron. Since the ash bat stores so little energy on impact, and returns that energy more efficiently than the energy disposed in distorting the ball, hardening the bat would seem to serve no very useful purpose. Indeed, if the hitting area were armored with a thin

---

[6]If you drop a ball and a bat, held vertically, an equal distance above concrete, the bat will bounce higher than the ball—and an aluminum bat will bounce much higher than a wooden bat.

sheet of absolutely hard and rigid steel, the well-hit ball would probably go slightly less far since energy would be lost as the distortion of the more elastic wood would be replaced by an increased distortion of the less elastic ball. But the scale of distance change is small at best. If the infinitely hard, armor-plated bat were substituted for the bats now in service, this would decrease the distance of a 400-foot home run only by about 2 feet. By and large, the benefit from any effort by players to harden their bats—legally or illegally—must be considered largely psychological.

The inelasticity is usually described in terms of a *coefficient of restitution* (COR), which is the ratio of the velocity of the ball rebounding from the surface of a hard, immovable object to the incident velocity and equal to the square root of the proportion of the energy retained in the collision. For baseballs traveling 85 ft/sec (58 mph), striking a wall of ash boards backed by concrete, the mean COR of a large set of 1985 and 1987 official major league baseballs has been measured to be 0.563; the balls rebound with a velocity of 0.563×85 ft/sec $= 48$ ft/sec. Since the energy of the ball is proportional to the square of the velocity, the rebound energy of the 85 ft/sec ball is about $0.563 \times 0.563 = 0.32$ times the incident energy; the collision is quite inelastic; 68 percent of the energy is lost to friction (satisfactorily close to the 65 percent found in the static measurements of Kirkpatrick).

My colleague R. C. Larsen augmented the results at 58 mph commissioned by Major League Baseball and measured the COR of one National League ball to be 0.588 for an initial ball velocity of 25 mph and 0.584 at 18 mph. Isaac Newton noted through his experiments that the COR is nearly independent of velocity for uniform spheres. L. J. Briggs, however, found that the COR of a golf ball fell off considerably with increased collision velocity, and the COR for baseballs probably also decreases as the collision velocity increases. In the course of his measurements at the National Bureau of Standards of baseballs for the army in 1942,

Briggs[7] found a value of COR of 0.46 at an incident velocity of 130 ft/sec (89 mph) for baseballs used in the major leagues in 1938. If these balls do not differ especially from today's balls, this indicates that the COR falls off sharply from the value of 0.563 measured at 58 mph to the value of 0.46 at 89 mph.

Of course, if the COR of the present ball is not much smaller at 89 mph than the measured values at 58 mph, the present ball might be assumed to be much livelier than the ball used in 1938. But this hardly seems likely; if the COR value of the present ball is as great as 0.55 for high-impact velocities, it would seem that the 360-foot fly ball Lou Gehrig hit in 1938 would go 420 feet into the bleachers[8] today and that the 58 home runs Hank Greenberg hit in 1938 were with a dead ball! (Figure 5.12 shows the variation of the distance a ball can be hit with respect to the COR.)

Assuming that Briggs's measurements apply to today's balls, we can make an educated guess as to the possible variation of the COR with velocity that is shown in the graph of Figure 4.4. These values are used in the following calculations of velocities— and then distances traveled—of batted balls.

While the COR values for balls hitting a flat wall are not, in general, the same as for balls hitting a cylindrical bat, the bat values are probably similar to the wall values. Hence, the wall-ball values of the COR plotted in Figure 5.4, used for the bat-ball collisions considered here, probably represent reality to a satisfactory approximation.

With the insight into the stress-strain characteristics of the

---

[7]Briggs, then Head of the National Bureau of Standards, published his results in the *NBS Journal* in 1945.

[8]The striking difference between the COR of the 1987 balls, measured to be 0.563 at impact velocities of 58 mph, and the COR values of 0.46, measured for the 1938 balls at an impact velocity of 89 mph, can be understood in one of three ways, all unattractive: (1) the values of the COR do not vary strongly with impact velocity and the 1987 balls are much more lively than the 1938 balls; (2) the balls are about the same but the COR falls off sharply between 58 mph and 89 mph; or (3) one of the measurements is wrong (if so, Briggs's 1942 measurement must be in error). We choose (2) as the least unlikely.

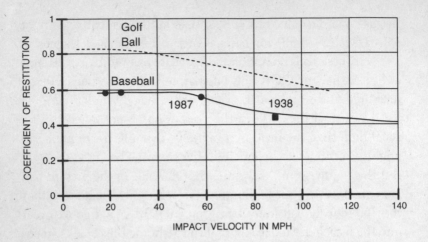

FIGURE 5.4 *Results of measurements of the coefficient of restitution (COR) of a 1987 baseball striking a rigid flat surface are shown by solid circles, and a value for a 1938 baseball is shown as a square. The solid line shows the estimate of the variation of the COR for present baseballs as a function of the impact velocity used in calculations. The broken curve shows the variation of the COR for a golf ball.*

baseball provided by Kirkpatrick's measurements shown in Figure 5.3, together with the information concerning the coefficients of restitution shown in Figure 5.4, we can use the simple physical principles of the conservation of energy and momentum to construct a fairly reliable description of the ball-bat impact. The relevant information is contained almost completely in the model stress-strain diagram of Figure 5.5. Note that the "distortion" in Figure 5.5 refers to the displacement of the surface relative to the center of mass and is nearly the distortion of the radius of the ball.

Using the information depicted in the graph, we derive the variation of force with time and the variation of ball compression with time[a] during the ball-bat collision shown in the graphs of Figure 5.6. Note that although the bat-ball contact extends over about $\frac{1}{1000}$ of a second—a millisecond—the time of strong impact lasts only about one-half millisecond.

The strength of the impact can be defined usefully as propor-

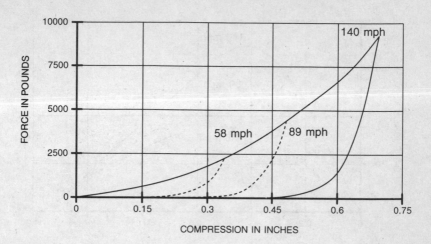

FIGURE 5.5 *Plausible stress-strain cycles for a baseball struck by a bat for impact velocities of 58, 89, and 140 mph. The velocities are in the center-of-mass system (as a ball striking a stationary wall); the displacements are with respect to the center of gravity of the ball and correspond approximately to changes in the radius of the ball.*

tional to the change in velocity of the ball. The velocity change corresponding to the impact velocity of 85 ft/sec (58 mph), at which major league baseballs are tested, is about 85 × 1.563 = 133 ft/sec, or 90 mph. This is about the same impact a baseball experiences when an 80-mph pitch is bunted—or when a stationary ball is given an initial velocity of 90 mph toward the outfield by a fungo bat. Briggs's measurements on the COR of 1938 baseballs at an impact velocity of 130 ft/sec (89 mph) concerned a velocity change of about 190 ft/sec, which is about the change that occurs when a 200-foot Texas leaguer is hit off a 65-mph curve ball. But when a 400-foot home run is hit off a 90-mph fast ball, the velocity change of about 290 ft/sec corresponds to an impact velocity of about 140 mph. It is hardly surprising that the maximum forces on the ball, and maximum ball distortions, are very much greater for the home run off the fast ball than the bunt.

The diagrams in Figure 5.7 provide a somewhat simplified

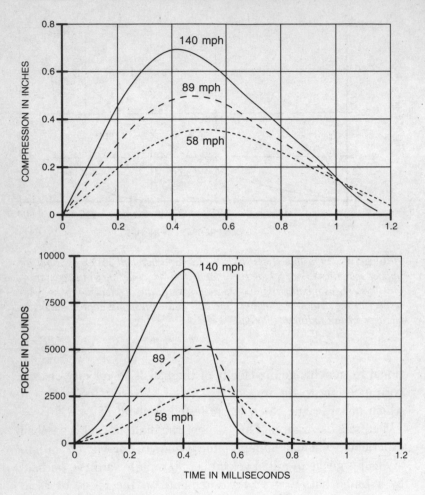

FIGURE 5.6 *The top figure shows the variation with time of ball compression in a ball-bat impact. The lower figure shows the variation with time of the force between the ball and bat.*

picture of the cross-section of the ball at maximum distortion for the three impact conditions considered. The ball's elasticity for the small distortions that occur for bunts—or fungo hitting—is largely determined by the character of the outer yarn windings and is not much affected by the properties of the center core of

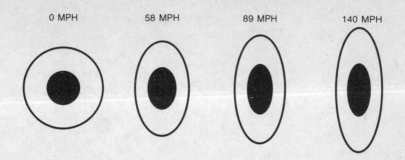

FIGURE 5.7 *Representative cross-sections of balls at the maximum compression during the bat-ball impact for various impact velocities. The true distortions will be similar but can be expected to show more flattening on the side that meets the bat.*

the ball. Conversely, the elasticity of the ball that is very much distorted by the impact that results in the home run off the fast ball is strongly affected by the character of the core of the ball. The major league tests of the elasticity at velocity reversals characteristic of bunts are useful, but they do not establish the constancy of the ball's elasticity for the long home run.

THE SPIN OF THE BATTED BALL

For almost all bat-ball collisions, the line of momentum transfer through the ball does not coincide with the final direction of the ball's flight; the bat does not strike the ball perfectly square. The ball is not lined straight back to the pitcher but is hit to left or right field, on the ground, or in the air. From such an off-line impact, the ball will leave the bat spinning, though it was not spinning originally. Figure 5.8 suggests the origin of the spin of the ball induced by oblique collisions with a stationary flat surface such as a wall. Here the ball, traveling with an initial velocity v, strikes the wall at an angle θ. The initial ball velocity can be divided into two parts, or components, as suggested by the dia-

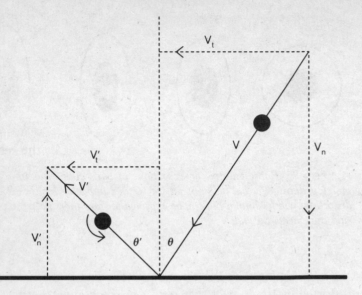

FIGURE 5.8   *Initial and rebound velocity components for a ball striking a flat surface at an angle. The curved arrow shows the direction of spin of the rebounding ball.*

gram: v_n is the component normal to the surface and v_t is the part tangential to the wall; v'_n and v'_t are the corresponding components of the rebound velocity v'.

We do not understand the stresses induced by the deformation of the complicated ball structure well enough to truly calculate the spin and trajectory of the ball as it rebounds from the wall, but we can deduce a sensible recipe that will allow us to make reasonable estimates of the state of the rebounding ball.

First, we can assume with confidence that the normal component of the rebound velocity is simply reduced from the incident normal velocity by a factor equal to the COR. For a value of the COR of 0.5, the normal velocity of the rebound v'_n would be just one-half of the initial normal velocity v_n. If the ball struck a very slick wall obliquely and skidded off it, the transverse velocity v_t would be unchanged by the collision. But the forces between the ball and wall are so large for the collisions that interest us that we expect the ball to *roll* on the wall in the course of the collision just as if its surface stuck there. Then the ball would come off the wall spinning, and some of its transverse energy would go into the spin energy, reducing the transverse velocity of rebound. Moreover, this rolling of the highly deformed ball must take up some additional energy in internal friction. We would expect that this energy loss would be less than, but related to, the dissipative energy loss that reduces the normal component of the rebound velocity. If the value of the COR were near one and there were no frictional reduction of the normal rebound velocity, we would expect no reduction of the tangential velocity from frictional losses.

For the normal component of the velocity, which is reversed on impact, if the COR is 0.5 the rebound will be one-half of the initial velocity. But the transverse component of the rebound velocity, which is not changed in direction, will retain about three-fourths of the initial transverse velocity.

We calculate the spin of the batted ball struck obliquely by the bat using recipes that follow these considerations[b]. Although these kinematic relations refer ostensibly to the collisions of the ball with a stationary flat surface, within the uncertainties of the recipes, they should adequately describe the collisions of the ball with the cylindrical bat when properly translated to the system of the moving bat.

Measurements by Briggs, cited in his *NBS Journal* report, that show that any spin of the incident ball, i.e., from the pitcher, is largely removed in the course of the bat-ball impact are substantiated by the observations of players who speak of "straightening out the curve ball." Hence, although the fast ball, with its backspin, will come off the bat with a little less backspin and at an angle about one degree lower than the same ball thrown without spin—and the curve ball, thrown with overspin, will come off the bat with a little more backspin and about one degree higher—we do not consider these small effects in our model. (A degree up or down represents a foot up or down as the ball passes the pitcher or two feet as the ball sails past second base.)

In the explicit calculations, we assume implicitly that the bat is swung approximately in a horizontal plane.[9] Then the ball pulled by a right-handed batter toward the left-field foul line will strike the bat at an angle of about 20° from normal and the ball will be spinning counterclockwise as seen from above as it leaves the bat. If the bat strikes the ball below the ball's center, the fly ball will take off with backspin; if the bat strikes the ball over the center, the ground ball that results will move across the infield with overspin. For a ball hit with a greatly tilted swing—such as the golflike swing at a pitch inside at the knees—the ball spin-axes are changed in an obvious fashion.

Since the spin of the ball affects its flight significantly, it is important to consider the spin induced by batting the ball in spite of the uncertainties in such spin calculations. Furthermore, even as there are uncertainties in our knowledge of the initial spin of batted balls, there are also uncertainties in our knowledge of the effect of the spin on the ball's flight (Chapter 2). Since the spin is important, however, we check the results of the models against that which we know from observations of the game.

As in pitching, the effects of up-down spin (usually backspin) are difficult to disentangle from the effects of gravity and then difficult to observe unequivocally. But the curves from sidespin are better defined. We thus calculate from our bat-ball recipe, and from our description of the Magnus effect in Chapter 2, the curve of a fly ball hit down the left-field foul line—over the left-field fence—and show the results in Figure 5.9. The ball carries an initial sidespin (clockwise as seen from below by the third baseman) of about 2000 rpm. We calculate that the ball, passing over the third baseman a few feet inside the base, curves about 25 feet foul at the 315-foot foul pole and lands in the street behind the ball park, after a carry of 380 feet, now "foul" by nearly 50 feet. Since this behavior is in rough accord with what we have all seen

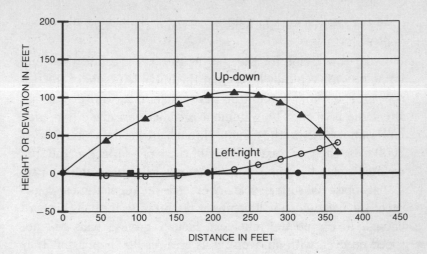

FIGURE 5.9 *The trajectory of a ball hit down the foul line showing the curvature produced by the sidespin induced by the bat.*

many times, we can conclude that our recipe is probably good within an uncertainty of, perhaps, 30 percent.

In these calculations, we assume that the motion of the bat's face is at right angles to that face and that all the sidespin of the ball follows from the ball's angle of incidence with the bat. Using the analogy of a golf shot, the club face is square with respect to its direction of motion. If this is the case, a ball hit to right or left field will always curve toward the foul line and a ball hit to center will have very little sidespin. Many batters, however, swing so the bat is moving along its axis when the ball is struck, giving a sidespin even to balls hit back toward the pitcher, in analogy to a slice or hook in golf. From watching the ball's flight, it seems that most batters hook a little—a right-handed batter hits the ball with a little extra counterclockwise spin (as seen from above), inciting a curve toward the left-field foul line. But some batters slice. Catcher Rich Gedman, who hits with an "open" stance (a slice stance in golf), with the forward foot farther from the plate than the rear foot, seems to slice the ball. Hitting left-handed,

Gedman's drives to right field may even curve toward the center fielder.

Using this recipe for the spin of the ball, we find that long home runs are typically hit so that the initial backspin of the ball is near 2000 rpm. Ground balls have similar rates of overspin that affect the bounce. The statement is sometimes made that such balls gain velocity as they bound through artificial turf infields. At 2000 rpm, the rotational velocity of the ball's surface at the spin equator is about 17 mph. A ball can only gain speed at a bounce, if that rotational surface velocity exceeds the regular linear velocity of the ball, which will be upward of 50 mph for balls hit hard enough to escape the infielders. Hence, ground balls will not speed up even with an overspin as great as 4000 rpm, but they will slow down less than balls with little overspin.

Rarely—but they are remembered in the nightmares of center fielders—line drives are hit to center field so that the ball has almost no spin, darts about like a knuckle ball, and is almost uncatchable. Balls hit to left or to right field will generally have enough sidespin to insure a smooth trajectory.

EFFECTS ON THE DISTANCE BALLS ARE HIT

Obviously, a ball travels farther if it is struck by a bat that is swung faster—or "harder." Less obviously, the fast ball hit by a full swing of the bat will travel faster and farther than a slow, change-of-pace pitch hit by the same speed swing. We will consider the magnitude of these effects.

For the sake of definiteness, in the following discussions we consider a bat, 35 inches long and weighing 32 ounces, striking a waist-high pitch so that the plane of the swing lies at 10° from the horizontal, i.e., the bat is swung slightly upward, to drive a ball at an angle of 35°. Among the variety of real swings, some vary considerably from this model swing; the swing at a ball passing over the inside corner of the plate at the knees often seems more like a golf swing. The salient points illustrated by the model, however, will be valid for all these swings.

Figure 5.10 shows the distance a ball that crosses the plate at a velocity of 85 mph travels when hit squarely with a full swing as a function of the bat speed. The assumption is made that the ball is hit at the center of percussion of the bat (a sweet-spot about $27\frac{1}{4}$ inches from the handle), and the quoted velocity is the velocity of that spot. Moreover, we assume that the ball is hit toward center field. A bat speed of about 77 mph is required to hit the ball 400 feet.

For a given bat speed, a solidly hit fast ball goes farther than a well-hit slow curve. Figure 5.11 shows the distance the ball travels for a bat being swung with a speed of 70 mph, striking balls that cross the plate at different speeds. The swing that hits a fungo (but with the bat described above, not a fungo bat) 300 feet will hit the 90-mph fast ball 370 feet!

Of course, a lively ball will travel farther than a dead ball.

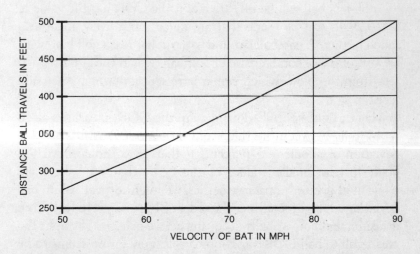

FIGURE 5.10   *The distance an 85-mph pitched ball can be hit by bats swung with different velocities.*

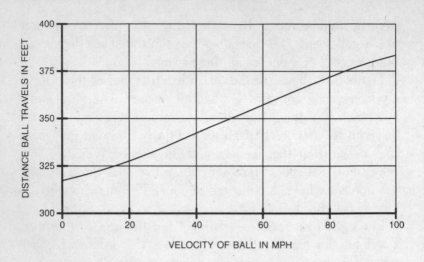

FIGURE 5.11   *The distance a pitched ball, traveling at different velocities, can be hit by a bat swung with a velocity of 70 mph.*

Figure 5.12 shows the distance a hard-hit home run ball will travel for different values of the COR under conditions such that the 90-mph fast ball will be hit 400 feet if the COR has the value of 0.430. Balls can be made that are much more lively and travel much farther. A baseball (wound with rubber like a golf ball) with a COR of 0.72 characteristic of golf balls would travel nearly 600 feet from the blow, which would have sent an ordinary ball 400 feet.

Major league specifications require the COR, measured at 58 mph, to lie between the values of 0.514 and 0.578, an allowable deviation of about ± 5 percent. If that proportional deviation holds for the smaller values of the COR that we believe are relevant at greater impact velocities, we might expect ball-to-ball deviations in flight length of the order of 10 percent. In fact, from measurements of 72 balls at an impact velocity of 58 mph, the mean ball-to-ball COR variation of balls now in use seems to be no larger than 0.005, corresponding to a deviation in distance of the 400-foot home run of only about four feet.

The sensitivity of the flight distance to the COR of a batted

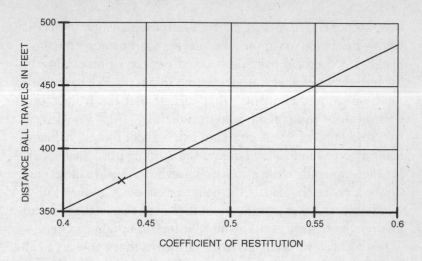

FIGURE 5.12   *The distance a batted ball can travel as a function of the value of the coefficient of restitution (COR) of the ball and bat if the bat velocity is 70 mph and the ball velocity is 85 mph. The* **X** *marks the estimated value of the COR for present balls and bats.*

ball is suggested by Figure 5.12. The graph shows the flight distance of the ball as a function of the COR for balls hit under conditions that would drive a ball 355 feet if the COR were correctly given by the graph in Figure 5.4.

In Chapter 2, we noted that the conditions of storage of balls could affect the elasticity of the balls and then the distance balls could be hit—long flies hit with balls stored under conditions of extreme humidity could be expected to fall as much as 30 feet short of the distance expected for normal balls. The elasticity of balls stored under extremes of cold or heat can be affected also. There are apocryphal stories, going all the way back to John McGraw when the home team supplied the balls to the umpire one-by-one, of home team managers storing the balls to be given the umpires when the visiting team was at bat on ice. Those balls were taken off of the ice a few hours before game time so the cover would warm up and not alert the umpire yet the core of the ball would remain cold and dead. Since McGraw managed, the rules have been changed so that *all* balls—still supplied by the home team—must be given to the umpires two hours before game time; hence, any doctored ball will be used with equal probability by home and visitors alike, rather evening things up. We have made some simple measurements[c] that suggest that the temperature effects can be substantial; deep-freezing to $-10°$ F would seem to take about 25 feet off of a nominal 375-foot fly ball, reducing many a home run to a loud out on the warning path. Such a ball, hit on the ground, would also skip through the infield a little more slowly. Conversely, storing the balls in a warming oven at 150° F would seem to inject enough rabbit in the ball to take the 375-foot fly over the fence to land about 400 feet from home plate.

In general, it seems that a swing of the bat that would hit a ball that was at a temperature of 70° F 375 feet would drive a ball an extra 3 feet for every 10° increase in the temperature of the ball. Conversely, the ball would travel about 3 feet less for every 10° decrease in the temperature of the ball. Even in very cold days, the ball temperature probably seldom drops very much below 60° since the ball in play was probably stored at room temperature before the game and will be kept from cooling too much by the handling of the pitcher. It is more likely that the ball will reach the ambient temperature on a really hot day; if the temperature on the field is 100°, the balls probably reach that temperature also.

Though there are still stories of teams that rely on singles, speed, and defense storing all balls in a deep freeze until a few hours before game time when they play a slower team that depends more upon power and home runs, I am not a believer. Since the careers—and salaries—of the home team batters aren't helped much by the lower batting averages that must result from hitting a cold dead ball, any cold treatment of the balls must be kept secret; impractically secret, to my mind.

It is interesting to note that a ball hit toward a foul line will go somewhat farther than a ball hit to dead center, though the bat speed is the same. This is because the ball must strike the bat at an angle (as shown in Figure 5.8) and the tangential part of the ball velocity v_t is not reversed by the bat-ball collision but retained with little loss in the blow. For the same bat speed, a batter will hit a 90-mph fast ball about 403 feet toward the left-center power alley with a swing that would hit the ball 400 feet to dead-center field. And if hit along the foul line, the ball would travel about 411 feet.

Table 5.1 lists a number of these effects and estimates of their effect on a standard 400-foot-long home run.

Although backspin adds distance to a long fly ball at a rate of about one foot in flight per 100 rpm of spin, there is probably little or nothing that a batter can do to change that spin. Any plausible batting stroke that takes a ball crossing the plate at a small downward angle of about 10° and propels it upward at the angle near 35° required to clear outfield fences will provide the ball with

Condition	Distance Added
1000 feet of altitude	+7 feet
10 degrees air temperature	+4 feet
10 degrees ball temperature	+4 feet
1 inch drop in barometer	+6 feet
1 mph following wind	+3 feet
Ball at 100 percent humidity	30 feet
Pitch, +5 mph	+3.5 feet
Hit along foul line	+11 feet
Aluminum bat	+30 feet

TABLE 5.1: *Distance added to 400-foot fly ball hit to center field.*

substantial backspin. Although the Ted Williams stroke that lofted Rip Sewell's Eephus pitch over the Fenway Park right-field fence in the 1946 All-Star Game probably generated almost no backspin, the Eephus pitch—tossed about 25 feet in the air to drop across the plate at a considerable angle—is seen today only in softball. Aside from losing the 20-foot flight advantage of backspin, Williams had to provide almost all the power—equal to an extra 50 feet or so of flight—to drive the 30-mph pitch over the fence. But only Babe Ruth could bat like Ted Williams, and no one else ever hit a home run off Sewell's blooper.

Batters sometimes work on their bats (illegally, but this is only a venial sin), grooving the surface to add backspin. It's unlikely, however, that there is any merit in this beyond occupational therapy. For all the small-angle impacts that lead to fair balls, the normal ball-bat friction is sufficient to ensure maximum backspin. Even as good, well-grooved tires do not stop a car faster than smooth, bald tires (as long as the car doesn't skid), fair balls hit off a smooth bat probably don't skid. Grooves, of course, are important on golf clubs, where hard steel, lubricated by the moisture from mashed grass, meets the smooth, enameled surface of the golf ball.

The inclination of the arc of the swing does not strongly affect the distance a ball is hit, but it does affect the mean angle at which the ball leaves the bat and, hence, the probability of hitting a very long ball and home run. The great high-average line-drive hitters such as Rod Carew and Wade Boggs swing the bat so that its barrel crosses the hitting region just in front of home plate, traveling upward on the same line that the average pitch is moving down, that is, at an angle of about 8 to 10°. (Players call this a level swing.) Then, if their bat position is correct but their timing is slightly off in an intent to hit the ball over second base, they will still hit the ball squarely. If they are early by 9 inches on the fast ball, as left-handed batters, they will pull the ball in a line drive between first and second into right field; if they are 9 inches late they will hit a line drive to the opposite (left) field. If their perfectly directed swing hits the ball absolutely square, that hit will be a line drive a few feet over the infielder's reach. If they hit the ball one-half inch high, they will hit a very hard ground ball that will often go through the infield; if they are one-half inch low, they will hit a high line drive for a double if it goes in the gap between fielders or down the line. But they will hit few home runs and very few pop flies. They will hit few home runs, not because they do not hit the ball hard, but because they do not hit the ball up. Boggs, in particular, hits the ball very hard and wins home run contests against certified sluggers by swinging up a bit on the ball.

The relatively low batting average but high slugging average home run hitters like Reggie Jackson uppercut the ball, swinging typically such that the bat is moving upward at an angle of as much as 20° as it meets the ball. Roughly speaking, every 10° extra, i.e., beyond 10°, of uppercut sends the squarely hit ball at an added upward angle of about 16°, so Jackson's squarely hit ball takes off, going up at about 26° and Carew's solid hit comes off his bat, moving up at an angle of 10°. But for equally hard-hit balls, Carew will end up sliding into second with a drive that landed about 210 feet from home plate and skipped between the out-fielders and Jackson will take a leisurely tour of the bases with the ball in the pocket of a fan who caught it 375 feet away in the right-field bleachers. If Jackson, the bigger man, actually hit the ball a little harder, it would go even farther.

But a mistimed swing will have different consequences for the two kinds of swing, as suggested by the diagrams of Figure 5.13. If they miss their timing by swinging 9 inches too early on the fast ball, Carew will still hit the ball squarely but now on a line over the first baseman's head. However, Jackson will top the ball and run out a weak grounder to the first baseman. If they swing equally late, Carew's squarely hit ball will pass in a line over the infield between short and third while Jackson, hitting under the ball, will hit a high pop fly to be gathered in by the third baseman. Hence, Reggie got the home runs while Rod hit with the high average. And both are on Bill James's list in his *Historical Baseball Abstract* of the 100 greatest who ever played the game.

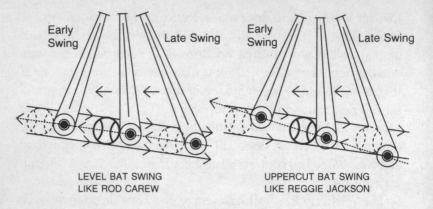

FIGURE 5.13   *Ball and bat trajectories showing the consequences of a mistimed swing by a left-handed batter with a level swing like Rod Carew and by a batter with an uppercut swing like Reggie Jackson.*

BATTING AGAINST THE FAST BALL

The examination of the mechanics of the ball and bat's trajectory suggests the difficulties that face batters such as Ernest Lawrence Thayer's immortal "Casey at the Bat" in his attempt to hit the ball safely and win the game for Mudville. To put the accuracy of stroke required to hit a baseball successfully in a human context, we will examine the character, the precision, and the timing of the decisions a batter needs in facing a fast-ball pitcher.

For each 1.5 mph over 90 mph, the ball travels an extra distance of about one foot past the plate in a given time—the ball is a foot quicker—and drops less (or, subjectively, rises more) by about one inch. If the right-handed batter, aiming to hit the ball to dead center, misjudges the timing and hits the ball squarely, but one foot late, the ball will sail down the right-field line foul—or if he swings $\frac{7}{1000}$ of a second early (a foot early) he will hit the ball foul down the left-field line. If the batter aiming to hit a line drive misjudges the drop (rise) by 1.0 inches, he will hit a long fly ball—perhaps even a home run! If he misjudges the rise by 2 inches, the extra rise from an incremental velocity of $2\frac{1}{2}$ mph, he will foul the ball back into the stands behind the catcher. And if he swings at the Todd Worrell 95-mph fast ball at the height he would expect a fast ball traveling (only) 91 mph, he will miss it altogether—even if he times it correctly.

Moreover, since the batter takes about one-fifth of a second to swing his bat, he must start the swing when the fast ball is about halfway to the plate. He still has a little time to change his mind and reorient his swing—but not much. After his swing is under-way for one-tenth of a second (and the ball is now about 15 feet from the plate), he won't be able to check it and he has little if any ability to change his point of aim. And about 50 percent of the deviation from curve, hop, or drop occurs in that last 15 feet.

And here we are considering hitting only the fast ball. If the batter sets himself for the fast ball on the three-and-one pitch, and the confident pitcher throws a curve or "change" instead with the same motion as the fast ball, the batter has serious adjustments to make in a very short time.[10]

In addition, to cover the 20-inch (17-inch plate width plus 3-inch ball diameter) strike zone, the batter must adjust his stride and make his decision about the direction of that stride and weight shift very early in his swing. It is therefore extremely difficult for a batter to prepare himself to hit a fast ball effectively that passes either on the inside or outside corner of the plate. Most batters must implicitly give up one corner or the other—if not both—before the ball is even pitched.

Batting is difficult, and the Mudville fans must try to forgive "the mighty Casey" for striking out. (With two strikes on him, I have always felt that Casey should have choked up a bit and simply tried to meet the ball rather than swing for the fence.)

---

[10]Warren Spahn said, "Hitting is timing." Then he added, "Pitching is upsetting timing."

LEFT-HANDED PLAYERS AND RIGHT-HANDED PLAYERS

Managers, players, and fans know that a right-handed batter has more trouble hitting against right-handed pitchers than lefties and left-handed batters have more trouble with left-handed pitchers. Late in the game, right-handed pinch hitters are chosen to face southpaw pitchers, and vice versa. Is this left-right effect real? If so, how big is it?

In Table 5.2, we summarize the results of four years' batting (1984–87) in the major leagues for all batters who had over 250 at bats (AB). The number of hits and the number of at bats are given for left-handed and right-handed pitchers while the batters are classified as left, right, or switch-hitters.

The switch-hitters hit .269 ± .0042 against left-handed pitching and .270 ± .0028 against right-handed pitchers[11] for an

---

[11]The quoted errors are estimates of "standard deviations." If only random chance variations affected the results, the "real" result would be expected to fall within the stated

Bat	Pitch					
	Left			Right		
	AB	Hits	Ave	AB	Hits	Ave
Left	25,121	6,451	.2568	69,188	19,756	.2855
Right	56,821	15,989	.2813	119,190	31,571	.2649
Switch	19,772	5,319	.2690	43,235	11,694	.2705
All	101,714	27,759	.2729	231,613	63,021	.2721

TABLE 5.2: *Batting statistics for 1984–1987.*

overall batting average of .2702. The left-handed batters hit the left-handed pitchers (LL) for a batting average of .2568 + .0036, while against right-handed pitchers (LR) they hit .2855 + .0023. Their overall batting average was .2778 + .00194 and their LR-LL differential was .0287 + .0043. Right-handed batters hit left-handed pitchers (RL) for a batting average of .2813 + .0052. Against right-handed pitchers (RR), they hit .2649 + .00168 for an overall average of .2702 + .00140. The RL-RR differential was .0164 + .0030.

For the average batter and average pitcher, changing pitchers (or batters) to take advantage of the left-right match is worth .0287 for left-handed hitters and .0164 for right-handed batters. For the average of all batters, the difference between the match and mismatch hitter-batter combinations is .020 + .0025; the average mismatched batter (RL or LR) has a 7 percent better chance of getting a hit than a matched (LL or RR) batter. So, don't replace a batter who matches the pitcher on left-right with a batter who mismatches but bats .020 points less. And don't replace a good mismatched pitcher with a poorer pitcher who matches. But we all know that. Or do we? And, of course, there are batters whose left-right difference is much greater—or much less—than average.

Why do lefties hit righties and righties hit lefties? The answer is not obvious. Of course, left-handed and right-handed pitchers throw from different angles, the lefty from the first-base side of the pitcher's box, and the righty from the third-base side. But the batter should have no trouble shifting his stance slightly—seldom more than 5°—to compensate for that angular shift.

Many, if not most, batters attribute the small but significant hitting difference solely to the different break of the curve ball. I have had players tell me that they do not care whether the fast ball is thrown left or right. But they find it more difficult to hit the curve that breaks out than the curve that breaks in. Right-

---

error about two-thirds of the time. The probability of being off by more than twice the error is about one in twenty.

handed screwball pitchers tend to throw screwballs to left-handed batters and curves to right-handed batters so that all batters, swinging from the left or the right, see balls that break away from them.

Why do batters hit in-curves better than out-curves? The out-curve that breaks over the plate is thrown at the batter. Some players say that such a curve freezes the batter briefly, thus reducing his adjustment time. But there are other reasons that might play a role. In the contest between pitcher and batter, the batter tends to undercorrect. In order to hit the variety of pitches that come his way, he must be poised to hit a medium-fast pitch over the center of the plate—and then adjust to the ball the pitcher actually throws. Against a good pitch, the batter tends not to adjust quite enough. Hence, batters hit under high pitches to send fly balls to the outfield, they hit over low pitches, bouncing them through the infield, and they swing late on fast pitches, driving them to the opposite field, and early on slow pitches, pulling them down the near foul line. They also hit over curve balls (which drop as much as curve) and under the hopping, rising fast ball. So when a pitch is such as to lead to compensating errors, the batter hits it squarely. Hence, the pitcher's fear of the "hanging curve ball," which is simply a curve thrown high in the strike zone. The batter tends to swing under the pitch because it is high, but over the curve ball because it is dropping; the over-under errors cancel so the ball is hit squarely and into the bleachers.

Such compensations may be important in hitting in-curves and out-curves. Batting against the curve ball, the batter tends to swing too quickly at the relatively slow pitch, and he tends to underestimate the in-out curve deviation. These errors tend to add up for out-curves but cancel for in-curves. Hence, the in-curves may be a little easier to hit, accounting for the small advantage batters have when they face a pitcher throwing from the opposite side. The diagram of Figure 5.14 illustrates this effect.

For a tailing fast ball thrown by a right-handed pitcher, breaking a little away from the batter, the right-handed batter's error in alignment and his late swing tend to add in a manner similar

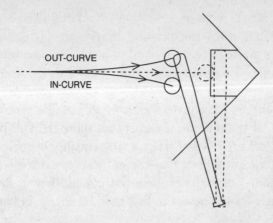

FIGURE 5.14 *The batter aims to hit a fast ball (dashed ball and bat) but the curve ball arrives later with a deflection (solid lines). The right-handed batter swinging too soon at the slower pitch hits the in-curve solidly, near the sweet-spot, but misses the out-curve or hits it weakly on the end of the bat.*

to that for the curve ball, while the errors compensate for a left-handed batter. Again, left-batter hits right-pitcher and right-batter hits left-pitcher best.

Left-handed hitters hit a little better, .0072 + .0024, than right-handed hitters. Nominally, left-handed batters have two advantages: (1) they are a step nearer first base, and (2) they face more opposite-handed pitchers (right-handed) who are presumably easier to hit against than same-handed pitchers (left-handed). Hence, we understand their higher batting averages. But do we?

Practically all who throw left-handed bat left (presidents of the United States, such as George Bush, excepted). Though many right-throwers bat left, about 70 percent bat right-handed, which suggests that there is some coordination or strength advantage to bat as you throw. The skill positions—second base, shortstop, third base, and catcher—where excellent fielding can compensate for mediocre batting, are reserved for right-throwers who tend to bat right. Southpaws *have* to hit well to play baseball field positions. Hence part of the L over R batting advantage must derive from the different positions available to left and right throwers.

An educated guess suggests that *all* of the left-right batting difference is due to the position difference. Hence, since the advantages are real, the left-handed hitters in baseball can be a little less gifted naturally than the right-handed hitters—as we expect.

Switch-hitters bat equally well from either side. From a casual inspection of the records, it seems that more than 90 percent of switch-hitters throw right. Hence, the equality of their left-right batting averages suggests that the natural advantages—strength? coordination?—of right-handers batting as they throw simply compensates for the nearer-to-first-base advantage of batting left-handed.

But there is one more factor: About 30 percent of the pitches thrown during the four years were thrown by southpaws. The near equivalence of total batting averages against left-handed (.2729) and right-handed (.2721) pitchers suggests that managers and general managers do a good job of selecting pitchers without left-right prejudice. But the proportion of left-handers in the male population, though not well known, is not likely to exceed 15 percent. Hence, left-handed pitchers are probably intrinsically slightly inferior to right-handers but are slightly favored by the asymmetric configuration of the game and the asymmetric distribution of left-right batting. Of course, this leaves open the possibility that left-handers are fundamentally superior to righties. As a southpaw myself, that's my position.

The remarkably complete and voluminous data on major league baseball players, including birth and death dates and whether they threw from the left or right, allows an analysis of left-right mortality that may have no parallel. This is especially interesting inasmuch as some have concluded—on the the basis of suspect data—that left-handed men do not live as long as right-handed men.

There is a problem of defining what it is to be left-handed. Those who are born southpaws, growing up in a right-handed world, have to learn to do many things with their right hand and their natural handedness is often blurred by social and environmental pressures. So how does one define a left-hander? Better not by the hand used to write or eat. I suggest that the hand used to throw a stone or ball is the most nearly culture-resistant index. Learned early, forgotten never, the left-hander always throws with his left hand. But what files record this handedness? Baseball records do, perhaps uniquely, and for a span of over 100 years. I take some of that data from *Total Baseball*[12] to see if left-handed and right-handed ball-players have different life-expectancies.

Table 5.3 shows the birth dates, death dates, and throwing handedness for 479 right-handed and 159 left-handed players selected randomly from the many thousands listed in the book. The majority of these players were not stars, of course (indeed, many came up for only a cup of coffee); but since we are studying longevity, not baseball excellence, this is not wholly relevant.

[12]*Total Baseball*, eds. J. Thorne and P. Palmer, Warner Books, New York. 1991.

	Right-Handed	Left-Handed
	Field Players	
Number	243	77
Birth date	1891.25±1.02	1895.1±1.82
Age at death	67.7±0.9	68.95±1.6
	Pitchers	
Number	238	82
Birth date	1891.9±1.04	1892.1±1.83
Age at death	64.7±0.9	65.75±1.55

TABLE 5.3: *Mean age at death and birth date for left-handed and right-handed field players and pitchers. The listed errors are standard deviations.*

About 13 percent of all field players and 25 percent of all pitchers were southpaws. Limited in numbers as it is, the set is sufficiently large so that the statistical uncertainties are matched by systematic effects; that is, a larger sample cannot be expected to give an appreciably more significant result from the limited information we have on the players.

The result (and the only significant result) is that *there is no significant difference between the longevities of left-handed and right-handed baseball players.* The left-handed players in the sample actually lived longer by one year, on the average, than the right-handed players, but that small difference was not outside the statistical and systematic uncertainties in the study. The mean difference of about three years between the lifetimes of pitchers and field players is suggestive of a truly shorter average life of pitchers, but that conclusion would have to be confirmed by a more careful study before acceptance. The mean birth date is important only inasmuch as it is about the same for the two classes of players, showing certain biases were not important. Likewise, the actual values of the mean lifetimes have no simple meaning. Nevertheless, the mean age at death of about sixty-seven for men born at a mean time of 1892 who lived to be about twenty-two and played baseball suggests that baseball players live as long or longer than the rest of the male population—be they left-handed or right-handed.

HOW FAR CAN A BALL BE HIT?

Stories of colossal clouts, of balls hit tremendous distances, constitute a part of the apocrypha of baseball. The stories of the longest blows (carries of from 550 to 600 feet are usually quoted) always place the action in some poorly surveyed situation, e.g., where the estimate pertains to where ball hit in the street outside the park and "estimate" is best translated as "wild guess."

Although the question as to how far a baseball can actually be hit is not easy to answer definitively, we can make some useful comments about the issue. It is surely not easy to hit a baseball 450 feet. But balls have been *caught* about that far from home plate in the deep center fields of the old Yankee Stadium and the Polo Grounds. At Yankee Stadium, Joe DiMaggio pulled down a Hank Greenberg drive in 1939 that traveled nearly 450 feet. And Willie Mays ran almost to the clubhouse stairs, about 475 feet from home plate in the old Polo Grounds, to catch a towering Vic Wertz smash with two men on to break the hearts of the Tigers in the 1954 World Series. With the hyperbole removed from a catch that didn't need it, the ball probably traveled about 450 feet.[13]

And there are home runs that we know have traveled nearly 500 feet where the distance is well defined simply by the direction and the fact that they were home runs. Phillip Lowry, in his book on ballparks, *Green Cathedrals*,[14] notes that Luke Easter (in a Negro League game in 1948), Joe Adcock, Lou Brock, and Henry Aaron all hit home runs over the 9-foot screen in front of the center-field bleachers 480 feet from home plate in the old New York Giants' Polo Grounds. On the side of limits, some interesting and well-defined negatives are known: *no one*—not Ruth, not Gehrig, nor DiMaggio, Mantle, or Maris—has ever hit a ball out of Yankee Stadium.[15]

We now have some interesting and accurate systematic information on the length of home runs from the IBM "Tale-of-the-Tape ©" program, which is designed to determine how far home runs hit by the home team traveled (or would have traveled if they had not landed in the stands). The scorer punches in the location where the ball landed (typically, the row and seat number, or where it hit walls, and screens) into an IBM personal computer loaded with coordinates from a survey of the park. The computer

---

[13]These balls would have stayed in the air about 6 seconds and that is time enough for a Mantle or Mays to run 150 feet under the ball.

[14]Addison-Wesley, New York, 1992.

[15]However, some say that Josh Gibson, the great catcher of the almost legendary Homestead Grays, hit a ball out.

finds the distance from home plate to where the ball landed and the height of the landing point above the playing field. Then the computer adds a length derived from that height to come up with the distance the ball would have traveled on a level playing field. The algorithms used are quite good, and the final distance is probably accurate to better than 10 feet.[16]

Figure 5.15 shows the frequency distribution of the length of the three longest home runs hit during 1988 and 1989 by home-team players from the teams which participated in the IBM program (15 in 1988 and 19 in 1989, of the then 26 major league teams.) Of about 2,000 home-team home runs, only two, a 478-foot blow by Dave Winfield and a 473-foot home run by Fred

---

[16]IBM makes a contribution to a charity designated by the team in the name of those who hit the longest home runs.

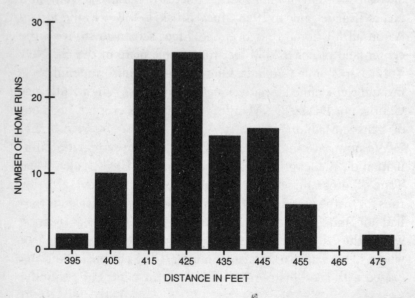

FIGURE 5.15 *The distribution of lengths for the three longest home runs hit at home by fifteen major league teams in 1988 and by nineteen teams in 1989. About 2,000 home runs were hit by the home teams.*

McGriff, carried over 460 feet; only eight, or about 1 in 250, traveled more than 450 feet.

So long home runs are surely rare, but if in two years there are two hits that traveled 475 feet, what is the record? Dan Valenti, in his book about famous home runs, *Clout,* [17] describes what might reasonably be considered the longest home run for which a reliable distance can be established. Mickey Mantle, batting right-handed, hit a Chuck Stobbs pitch out of Griffith Stadium in 1953 that was claimed to have traveled 565 feet by a Yankee publicity man. Mantle's blow surely didn't go that far, but it did go a long way and we can estimate how far, rather accurately.

The ball was seen to glance off of a beer sign in left-center field as it cleared the 55-foot left-field bleachers 460 feet from home plate; the place it hit the beer sign was 60 feet above the playing field level. In general, balls go farthest when hit at a launching angle of about 35°; with a strong following wind, the optimum angle is more like 40°. And—as a consequence of air resistance—a ball comes down at an angle greater than that from which it takes off. If the ball came down at an angle of 50°, it would have hit the ground 510 feet from home plate; a more precise calculation gives an answer of 506 feet with an uncertainty which I put as no more than 5 feet.

A probable trajectory of the ball Mantle hit is shown in Figure 5.16. It is obvious from the scale drawing that the claim of 565 feet is nonsense, but that an estimate of 506 feet makes good sense.

When Mantle hit Stobbs's pitch, there seems to have been a strong following wind—I expect that the longest home runs hit in outdoor parks are always wind-assisted. Valenti quotes Sam Diaz, a meteorologist who worked that day in the Washington Weather Bureau, to the effect that the wind during that period was 20 mph with gusts up to 41 mph. If we take the wind as blowing out at 20 mph 60 feet above the playing field, but

---

[17]Stephen Greene, New York, 1989.

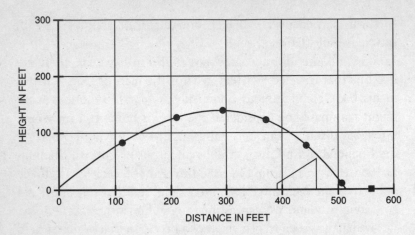

FIGURE 5.16   *A plausible trajectory of the ball hit by Mickey Mantle off of Chuck Stobbs that went over the left-field bleachers at Griffith Stadium. The solid square shows the distance the ball was claimed to go. The circles show the position of the ball at intervals of one second.*

shielded by the stands below 60 feet, we can conclude that a ball hit that hard would carry about 430 feet on a windless day; and if the wind had been against Mantle, the ball would have been only a routine fly-ball out. However, major league baseball had been played in Griffith Stadium for almost fifty years, and the wind must have been blowing out often, but since no one else ever hit a ball out during a major league game,[18] Mantle's feat shouldn't be denigrated.

Although the question as to how far a baseball can actually be hit must eventually be answered through experiment, it is possible to make some useful comments on the question. First, though many balls are hit 400 feet, such a blow constitutes quite a long home run. According to the graph in Figure 5.10, the required bat speed required to hit an 85-mph fast ball 400 feet to dead center field under standard conditions (at sea level, no wind,

---

[18]Josh Gibson is said to have cleared that wall *twice* when the Homestead Grays played home games at Griffith Stadium in the 1940s.

moderate temperature, and so forth) would be about 76 mph. To hit the ball 450 feet, the bat must be swung at a velocity of 86 mph, 13 percent faster with an energy 28 percent greater; and to go to 500 feet, 50 percent more energy is needed. Since this is a very large difference, we then estimate that 450 feet is about the maximum that ball players can hit the ball *under such standard conditions.*

Although 450 feet may be the maximum distance a ball can be hit under *normal* conditions in Old New York, baseball is played under other conditions and in other places. Let us say that the pitcher is very fast and the ball velocity is then 95 mph. This increases the distance the ball would be hit from 450 to about 455 feet. If the ball is pulled toward the foul lines, it will go as much as 15 feet farther and we are up to 470 feet. Let us assume further that we are discussing a game played on a 100° day in July; then the ball could go 10 feet farther because the hot air is thinner, and another 10 feet farther because the hot ball is livelier, to land 490 feet from home plate. If a 10-mph following wind aids the flight, the ball could go another 40 feet and we have a 530-foot blow. (We exclude meteorological freaks; the tornado that wafted Dorothy from Kansas to Oz could well have taken a baseball along.) In the thinner air of Atlanta—perhaps as the barometer is falling with lightning on the horizon—the ball might go another 10 feet to give us 540 feet, and in Denver, 570 feet! And if the foreman in the ball manufacturing factory in Haiti had set the tension on the winding machine too high one afternoon so as to generate a gross of rabbit balls, who is to say how far one of *those* balls might go off of a strong hitter's bat?

In summary, if someone tells me that a ball was hit 550 feet anywhere in the majors but Denver, I won't believe it—but I won't bet the farm against it. If the ball *is* hit in Denver, it might nearly reach Kansas, or even Oz.

In my estimates of maximum-length home runs, I have limited myself to home runs by *major league baseball players in games.* If a sufficient prize were given for the ball hit the farthest under baseball rules, one might recruit an Olympic-level weight man, quick as a cat, and carrying nearly 300 pounds on a near-seven-foot frame, to do the job. Swinging a 42-inch-long, 4-pound bat by taking two steps—like the young Babe Ruth—into fast balls thrown over the center of the plate, I estimate that he could hit the ball well over 500 feet under standard conditions at sea level.

If a ball player can hit a ball 550 feet, how far can one throw a ball? In 1957, at Omaha, Glen Gorbous threw a ball 446 feet. Assuming he threw on a hot day with a 10-mph following breeze at the Omaha altitude of 1040 feet, we estimate that Gorbous must have thrown the ball with an initial velocity of about 120 mph. In Chapter 3, we noted that the very fastest pitchers throw with a "muzzle velocity" of about 110 mph. But pitchers throw from a stationary position while Gorbous threw from a run, which is sufficient to account for the extra velocity.

TECHNICAL NOTES

a. For a given velocity of impact, the area under the upper curve in Figure 5.5 is proportional to the initial impact energy, the area under the lower curve is proportional to the energy returned to the rebounding bat and ball, and the area enclosed by the curves is proportional to the energy dissipated in the distortion of the ball. The area under the force-time curve of Figure 5.6 is equal to the change in the velocity of the ball in appropriate units.

b. We use the following formulae in the course of calculations of the batted balls' trajectories:

$$v'_n = Cv_n, \quad v'_t = v_t \times \sqrt{\tfrac{5}{7}} \times \sqrt{0.5C + 0.5} \quad \text{and} \quad \omega = \frac{v'_t}{R}$$

where C is the COR, v_n and v'_n are initial and rebound normal velocities in the ball-bat center-of-mass system (which is equivalent to the ball-wall system of Figure 5.8), v_t and v'_t are the associated tangential velocities, ω is 2π times rate of spin of the ball (in revolutions per second), and R is the ball radius. The factor $\sqrt{\tfrac{5}{7}}$ accounts for the transfer of translational kinetic energy to rotational energy, and the factor $\sqrt{0.5C + 0.5}$ is introduced as an estimate of the reduction of the transverse rebound velocity that might follow from frictional effects. These recipes are designed to represent collisions at an angle less than 30° from the normal (which includes all fair hits). In the case of more nearly glancing collisions, the ball may skid off the bat, losing little transverse velocity and picking up little spin.

c. My colleague R. C. Larsen put several balls in a deep freeze held at about −10° F and several in an oven held at about 175° F for about a day. He then let the balls sit at room temperature for one hour and measured their coefficient of restitution (COR) at a velocity of 25 mph. As one might guess, the cold balls were dead—their COR was reduced about 10 percent from normal—and the heated balls were quite lively— their COR increased about 13 percent. This translates to a change in the distance of a fly ball hit 375 feet with a normal ball to about 350 feet for cold balls and over 400 feet for the hot balls.

PROPERTIES OF BATS

THE OPTIMUM BAT WEIGHT

From the *Official Baseball Rules:*

1.09 (a) The bat shall be a smooth, rounded stick not more than $2\frac{3}{4}$ inches in diameter at the thickest part and not more than 42 inches in length. The bat shall be

    (1) one piece of solid wood, or
    (2) formed from a block of wood consisting of two or more pieces of wood bonded together[1]. . .

(b) Cupped Bats. An indentation in the end of the bat up to one inch in depth is permitted and may be no wider than two inches and no less than one inch in diameter. The indentation must be curved with no foreign substance added.

(c) The bat handle, for not more than 18 inches from the end, may be covered or treated with any material to improve the grip.

In fact, the bats used effectively by players are almost never longer than 36 inches. Once made of hickory and weighing from

---

[1]Any laminated bat design must be approved by the Rules Committee. In practice, laminated bats are not used.

40 ounces on up to the 56 ounces of lumber[2] supposedly wielded by the young Ruth, today's bats are made of ash and range mostly from 31 to 36 ounces. The typical dugout bat rack in 1920 probably held no bat lighter than 36 ounces; the 1989 bat rack probably holds no bat heavier than 36 ounces. But then and now, the choice of weights and shapes varies considerably from player to player.

If we understand the mechanics of batting, it would seem we could establish the optimum size and weight of the bat from simple physical principles. A number of factors, however, suggest that it may not be possible to reach definite conclusions about the best size and weight for a bat: the broad choice of bats used successfully by baseball players tells us that proficiency in batting cannot depend sensitively on the bat's character; the game's sensitivity to small differences in batting means that precise analyses may be required if reliable conclusions are sought concerning those small differences (there is only about a 5 percent difference in the length of flight of the average home run and a long fly ball caught on the warning path), and the act of batting is, in itself, complex and not precisely understood.

At this time we will describe our analyses solely in terms of an optimum weight of the bat since this—along with the bat length—is a commonplace parameter. We neglect differences in the distribution of the weight along the bat, i.e., the shape of the bat, though that can be important; obviously, the addition of an ounce in the bat's handle has much less effect on the swing than the same ounce added to the barrel. We emphasize home runs because home runs are important and less dependent on tactical decisions than lesser blows.

The simplifying assumption that the weight of a bat can be changed without modifying its shape and weight distribution can only be exactly valid if a complete range of wood densities is available—and that is not the case. Baseball bats seem to have

---

[2]Hack Miller, Cub outfielder of the early 1920s, was said to have swung a 69-ounce bat!

been made only of hickory or ash; the heavy bats used largely in the first quarter of this century were made of hickory with a density of about 0.82 times that of water; the light bats used today are turned from ash with a density of about 0.638. A hickory bat with the same dimensions as the 33-ounce ash bat Roger Maris used to hit 61 home runs would weigh about 42 ounces. The 48-ounce hickory bat Edd Roush swung (choked up) to give him a lifetime batting average of .323 from 1913 to 1931 and the 47-ounce hickory bat Babe Ruth swung (held at the end) to hit 60 home runs in 154 games in 1927 (a record that still stands) had larger barrels and thicker handles than Maris's bat and were probably an inch or so longer.

The rules of baseball allow a barrel diameter of 2.75 inches. In order to keep the bat weight down, the light bats used today generally have smaller barrels—typically 2.50 inches in diameter, which reduces the bat-ball hitting area (which is proportional to the sum of the diameters of ball and bat) by about 4 percent and the weight by about 4 ounces for an ash bat. The heavy bats Mickey Mantle used and the 40-ounce bat Dick Allen swung had full 2.75-inch barrels. Of course, longer bats weigh more; an inch of extra length in the barrel adds nearly 3 ounces to the weight of a hickory bat and a little more than 2 ounces to an ash bat, though an inch added to the handle adds only about a third of an ounce.

The great home run hitters have used a broad spectrum of bat weights. Early in his career, when he led the American League with 11 home runs in 1918, though he was used mainly as a pitcher and spent only about one-third of the season in the outfield, Ruth was reputed to have sometimes used a 56-ounce hickory bat. In 1927, when he hit 60 home runs, he seems to have favored a 47-ounce Hillerich & Bradsby slugger. Later he dropped to a 44-ounce bat, and was said to be experimenting with bats as light as 36 ounces in his last year, 1935, with the Boston Braves— when he quit early in the season with only 13 hits in 72 at bats, though six of the hits were home runs. And Ruth's last home run, the last he ever hit, was the first ball ever hit over the right-field stands at Forbes Field in Pittsburgh!

But most players used much lighter bats then and use them now. Roger Maris hit 61 home runs with a 33-ounce bat, and Hank Aaron used 31- and 32-ounce bats to hit more major league home runs than anyone else. Though Ernie Lombardi used a 42-ounce bat in the 1930s and 1940s, no other major batter since Ruth seems to have hefted more than 40 ounces.

If one player can hit a significant number of home runs with a 56-ounce bat and another player can be equally successful with a 32-ounce bat, one is led, inevitably, to the conclusion that the distance a well-hit ball travels cannot depend dramatically on the weight of the bat. Furthermore, since there are obvious advantages to a light bat, as it must be easier for a batter to time a pitch with the quick swing possible from a light bat, there must be compensating advantages to a heavy bat or Ruth and Lombardi would not have used heavy bats. Though the heavier, and larger, bat, with its larger and longer barrel, provides more good hitting area, such a bat probably also transfers a higher velocity to the well-struck ball. It seems likely that a player can actually hit the ball a little farther with a heavy bat.

But could the players choose badly? Could an analysis of the physical principles of batting lead us in a different direction? Because of the difficulty of obtaining a full understanding of so complex an activity as batting, we must begin simply. From elementary principles of mechanics, we can say with complete reliability that for a given bat speed a heavy bat will drive a ball farther than a light bat. Conversely, for a given kinetic energy of the bat, a light bat will drive a ball farther than a heavy bat (for bat weights greater than 20 ounces). To this we add the (very plausible) condition that no player can swing a heavy bat faster than a light bat. And we hold that no player can put more energy into a light bat than a heavy bat. The energy transmitted to the bat is simply the product (better, *integral*) of the force applied to the bat through the hands of the batter times the distance through which his hands move. If we assume that the arc of the hands—and bat—is the same for light and heavy bats, the larger force that it is possible to apply to the more slowly moving heavy bat will result in a larger energy transfer to the heavy bat.

Figure 6.1 shows the variation of maximum distances with bat weights that a 90-mph fast ball can be hit under these two conditions; the broken line shows the variation if the different-weight bats are swung with the same kinetic energy that drives a ball 381 feet from a 32-ounce bat; the dotted line shows the variation if the bats are swung with the same velocity as that of the 32-ounce bat. Noting that these curves establish *limits* on the variation of distance with bat weight, we can be reasonably confident that if a man who hits a ball 381 feet with a 32-ounce bat changes bat weight, and swings just as hard, the distance the ball will go will fall between these two extreme lines. In addition, it is clear that the man who hits 381-foot home runs with a 32-ounce bat is not likely to gain or lose very much distance by changing the weight of his bat by a few ounces.

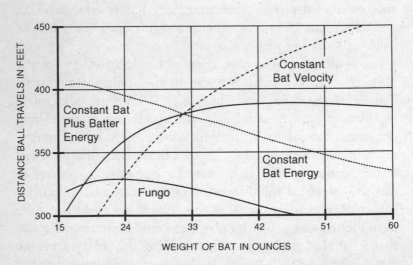

FIGURE 6.1   *The curves show the distances a 90-mph fast ball can be hit by bats of different weights swung at a definite velocity, swung with a definite energy, and swung so that the total bat-batter energy is held constant when a 90-mph ball is hit and when a stationary ball is hit (a fungo).*

Although this is as far as physics can carry us with—almost—complete reliability, it is interesting to investigate more detailed results from a simple, but plausible, model of bat and batter. In this model, we assume that both the bat and the body of the player move in the same general path, driven by the same muscular forces, though different-weight bats are used[a]. The forces generated by the muscles move the 180-pound batter just as they move the 2-pound bat. Since the same forces are applied over the same distances, the total energy generated to move bat and body during the swing is the same for different bat weights. If the bat is light, man and bat will move a little faster; if the bat is heavy, the swing will be a little slower. Fast or slow, light bat or heavy, the total kinetic energy of man and bat will be held constant, as we assume a constant muscular force and a constant path of motion through the swing. Though the energy is held constant for different bat weights, it is divided differently between bat and body. More of the energy goes into a heavy bat, though the bat—and player—velocity is smaller; less energy goes into a light bat, though the velocity of bat and player is somewhat greater. In the case illustrated by the solid line in Figure 6.1, the kinetic energy of the 32-ounce bat is about the same as the kinetic energy of the 180-pound ballplayer. We again choose the total energy as that which would drive a 90-mph fast ball an optimum distance of 381 feet.

The real batter-bat interaction is much more complicated than the model. In fact, energy is exchanged between bat and batter in an intricate fashion. Moreover, physiologists note that the forces put out by muscles decrease as the velocity of contraction of the muscles increases, suggesting that less force is applied generating less energy when the muscles move more quickly in swinging the lighter bat. Hence, the model is not meant to describe the batter-bat interaction in detail but to simulate the results of the real interaction in some useful approximation.

The model is certainly simplistic, and the conclusions derived from it must be taken lightly. But the results that suggest the maximum distance of 389 feet would be achieved if the batter would use a 46-ounce bat are interesting and not implausible. The effects are not large: the 46-ounce bat propels the ball only 8 feet farther than the 32-ounce bat; if the batter drills a hole in his 32-ounce bat (perhaps filling it with cork) so the weight is reduced to 28 ounces, the ball will go only 373 feet.

Armed with the model (albeit simplistic), one can look at other factors. How does the optimum bat weight vary with the size of the player? Babe Ruth was reported to weigh 251 pounds in 1927 when he hit 60 home runs. According to our model, the Babe would have been best suited to hit 90-mph fast balls for distance with a 52-ounce bat. Conversely, Lloyd Waner probably weighed little more than 140 pounds at times. Our model tells us that "Little Poison" would hit fast balls hardest with a 42-ounce bat.

Using the approximation that the total energy output of a ballplayer is proportional to his weight, we can estimate the dependence of the length a player can hit a ball on his size. Hence, we find that the same proportionate effort that allowed the 140-pound Waner to drive a 90-mph fast ball 338 feet with a 32-ounce bat would allow 180-pound Ducky Medwick to hit the ball 362 feet using the same-weight bat. And if 225-pound Mark McGwire swung equally hard, his shot could land in the bleachers 382 feet from home plate.

If the ball is moving more slowly, the optimum bat weight declines. The 180-pound nonpareil model player would best use a 48-ounce bat to reverse the velocity of a 95-mph Roger Clemens fast ball and send it toward the center-field fence. But a 42.5-ounce bat would better drive the 80-mph slider, and the 65-mph breaking ball would be hit farthest by a 37-ounce bat. And if the batter wanted to hit fungoes (or bat a ball off a tee), he would find that a 24-ounce bat would be best—if he could find one. Indeed, special fungo bats are made to be quite light by reducing the diameter of the barrel.

By and large, these results are in accord with experience and common sense. Except for the fungo bat, however, all of the optimum bat weights seem rather high, even if they are optimized for slower ball velocities than the major league fast ball. But to survive in professional baseball, one must hit the ball often as well as hard. The model player would lose only a little distance by going to a 32-ounce bat, and with the extra quickness of swinging the lighter bat, he could expect to make good contact more often.

Have any measurements been made that might substantiate the results of the simple model that heavy bats drive the ball farther? Yes! Many years ago, *This Week* magazine (May 20, 1962) reported the results of an experiment in which Roger Maris batted for distance with five different bats that varied from 33 ounces to 47 ounces. The bats were copies from Hillerich & Bradsby records of the bats of previous great home run hitters (Ruth used the 47-ounce bat to hit his 60 home runs). The pitcher was a veteran Yankee batting-practice pitcher who expertly served up hittable pitches. Though there were fluctuations in the length of the drives, of course, a statistical analysis confirmed the trend seen casually: the heavier the bat, the farther Maris hit the ball. Indeed, the best fit to the data showed an increase of range with bat weight about like that given by the dotted constant-velocity line of Figure 6.1, though that limit was within the uncertainties of the data. Maris probably adjusted his swing so that he swung the heavier bats almost as fast as the lighter bats; perhaps he lengthened his swing when he used the heavier bats to hit the batting-practice pitching.

However, although the average length of balls he hit with his own 33-ounce bat was the shortest of all, Maris returned to that bat to face hostile pitching.

If the loss in distance that follows from a choice of a lighter bat is not great, is the gain in time in swinging the lighter bat substantial? It is easy to set limits on the difference in timing using the extreme model of the constant-energy bat. If we assume that the "time of decision" is typically 0.15 seconds before the ball crosses the plate, we can conclude with some assurance that the gain in the time required to swing the bat that follows from the change from a very heavy 38-ounce bat to a moderately light 32-ounce bat can be no greater than 0.0133 seconds, or about 1'9", in the flight of the 90-mph fast ball. Our more realistic model cuts that down to about 1'3". The batter would gain a little over a foot on the fast ball by shifting from a 38-ounce bludgeon to his favorite 32-ounce bat. If he drills a hole in that 32-ounce bat, reducing the weight in the barrel by about $1\frac{1}{2}$ ounces, he would gain about 0.005 seconds, or about 6 inches, on the fast ball. For some players, the small gain in time may be more important than the small loss in the speed—and flight distance— of the ball as it comes off the slightly lighter bat.

Such a simplistic, essentially kinematic, argument ignores the athletic complexity of the swing of the bat. The fine batter's "beautiful" swing that supplies maximum energy to the bat as it strikes the ball follows from precise timing of the various actions that contribute to the swing. It seems possible that for a given batter, a slightly lighter—or slightly heavier—bat may lead to a better (timed) swing and a larger energy transfer to the bat. And this could well override any pedantic kinematic derivation of an ideal bat weight.

Though these discussions define the considerations applicable to choosing an optimum bat weight—and support the common-sense view that players now choose bat weights sensibly—is it not possible that a better weight distribution would serve to drive a ball faster and farther? Could the standard bat be badly designed?

The kinetic energy the player supplies to the bat can be divided into two parts: an energy associated with the linear motion toward the ball by the mass of the bat and an energy associated with the bat's rotation about the point of impact with the ball. This rotational energy, which is typically about 5 percent of the total energy, does not contribute toward driving the ball. If a bat could be designed so that almost all of the mass were concentrated very near the point of impact (a bat with a short, heavy barrel and a handle like the wooden shafts of Bobby Jones's golf clubs of the 1920s would do the trick), the rotational portion of the bat energy would be much reduced. Such a bat, swung with the same total energy, would hit the ball somewhat farther; the ball hit 400 feet with the standard bat would go almost 410 feet from the golf club–bat. But the small increment of extra distance would be achieved at the cost of a considerable reduction of the region of good hitting and a considerable increase in the cost of broken bats. A ball hit a few inches from the optimum point along the axis of the bat would probably not leave the batting box much faster than the broken-off head of the bat.

HITTING THE BALL IMPERFECTLY

This discussion of batting has so far only considered almost perfectly hit balls. Since it is the pitcher's goal to see that balls are not well hit—and pitchers are sufficiently successful to hold mean batting averages well below 30 percent—mis-hit balls are the rule rather than the exception. The bat must be swung with great accuracy with respect to the trajectory of the pitched ball if the ball is to be well hit. We have noted that the swing must be timed correctly with an error no larger than $\frac{1}{100}$ of a second if the ball is to be hit fairly between first and third base.

Rather precise vertical orientation is also required. Figure 6.2 shows the trajectories of balls hit by a bat with a 2.75-inch-diameter barrel with different bat-ball vertical displacements. We assume that the swing is otherwise well directed and the bat is swung with a speed of 70 mph upward at an angle of 10° at an 85-mph overhand fast ball, which crosses the plate dropping at an angle near 10°. The ball that is hit squarely is driven 200 feet as a line drive, rising to a maximum height of about 13 feet. If the bat is swung under the ball by 2 inches so that the line of motion of the bat's center passes under the ball's center by 2 inches, the ball will go onto the roof of the stands behind the catcher. The ball hit one inch below center will be a routine fly ball out unless it is hit down the foul line, where it might be a home run. The ball will go farthest—about 380 feet—if hit about $\frac{1}{2}$ of an inch below center. If the bat has a smaller 2.50-inch barrel, as is the case for most of the light bats used today, the characteristic offsets must be reduced by about 4 percent.

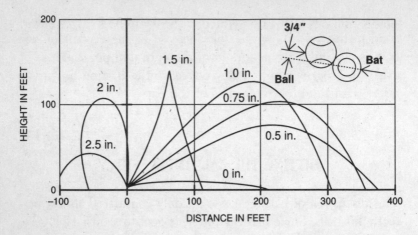

FIGURE 6.2   *The trajectories of balls struck by a bat swung under the ball by given amounts. The bat speed is 70 mph directed upward by 10°; the ball speed is 85 mph directed down by 10°. The diagram in the upper right shows the configuration of the bat and ball when the bat is swung under the ball by three-fourths of an inch.*

The paths of ground balls are more difficult to show, but, excepting Baltimore chops, where the ball rebounds from home plate so high that the batter is on first before it comes down, the ground ball that goes for a base hit must not be hit too sharply into the ground. By and large, for a ball to be hit so that it evades the infielders, the bat should strike the ball no lower than $\frac{3}{4}$ of an inch below center and no higher than $\frac{3}{8}$ of an inch above center.

The batter is permitted a larger error along the bat's axis. An error of 2 inches is quite permissible, and if other criteria are satisfied, he can hit the ball well, though the impact is more than 3 inches along the axis from the optimum point on the bat.

Though the nominal latitude of error of the position of the ball-bat impact point along the bat's axis is greater than the allowable up-down error, this batter can adjust to changes in the ball's vertical position more easily than changes in its horizontal position. He does not need to change his stride or body orientation sharply to deal with the allowed vertical range of strike pitches, but he must change the position of his body to cover the 20-inch width of the strike zone. (The strike zone is one ball diameter wider than the plate, since the umpire is supposed to call a pitch a strike if any part of the ball passes through the strike zone.)

If a bat strikes a ball at a point on the bat at a distance from an optimum point too near the handle or end of the bat, the velocity of the ball from the bat is reduced. For expositional simplicity, it is interesting to describe this loss of velocity in terms of a loss of distance in a long drive to the outfield. The velocity variation with position along the bat axis is determined by simple kinematic criteria, defining the loss of energy from collision-induced rotation of the bat and the loss of energy to vibrations of the bat along its length.

The broken curve in Figure 6.3 shows the variation of distance the ball travels as the point of impact varies along the bat if the bat is (unrealistically) so stiff that oscillations can be neglected and (unrealistically) one can neglect the clamping of the end of the bat by the hands of the batter.[b] For the particular bat under

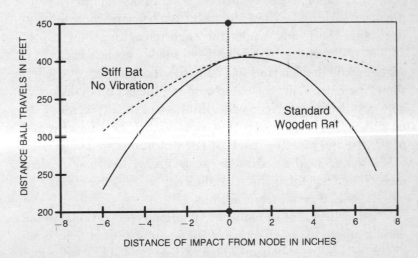

FIGURE 6.3 *The variation of the distance a ball can be hit as a function of the distance of bat-ball impact from a vibrational nodal point of the bat.*

consideration, which is 35 inches long, the zero point is chosen to be a point (the center of percussion) about $27\frac{3}{8}$ inches from the handle. The bat handle will be driven forward into the left palm of the right-handed hitter by bat-ball impacts toward the end of the bat beyond the center of percussion; the handle will be driven backward into the right palm by impacts nearer the handle; impacts at the center of percussion will not affect the motion of the handle. The force exerted on the hands by a mis-hit with a stiff bat is not great, not nearly enough to sting the hands. The ball's impact—which would sting the bare hands if caught—is greatly reduced by transmission through the bat. The maximum energy transfer from the stiff bat to the ball occurs when the ball is struck at a point about 30 inches from the handle or about 5 inches from the end of the bat. Energy is lost in the impact of balls hit at the end of the bat, or near the handle, by the increased rotational energy the impact transfers to the bat. Hence, the variation in distance, reflecting a variation in velocity of the driven ball, follows simply from kinematic criteria.

But bats vibrate when the ball is hit too far from the optimal point, a vibrational node (which is a point of no vibration), resulting in a weakly hit ball and often a broken bat. And that vibration does sting the hands. Under the large forces the ball's impact applies to the bat, a bat's vibrational mode becomes excited, as suggested by the diagrams in Figure 6.4. The diagram at the left shows the bat striking the ball at a node[c]; the bat is not distorted. The next two diagrams show the distortion of the bat (center-line) on striking the ball at points of maximum vibration—near the end of the bat and near the handle. (The distortions are exaggerated in the diagrams; the maximum excursions are less than one-half inch.) The diagram at the far right suggests the envelope of the vibrations that result from mis-hits.

The magnitude of the induced vibration is proportional to the natural amplitude of vibration at the point of impact; hence, no vibrations are set up when the ball strikes the bat at a node and maximal vibrations are set up if the point of impact is near an anti-node, a point of maximum vibration amplitude. The energy

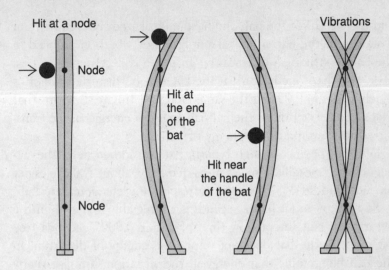

FIGURE 6.4 *Distortions of a bat on striking a ball at a node, at the antinode at the end of the bat, and at the antinode near the center of the bat. The diagram at the right suggests the character of the vibrations induced by hits away from the node.*

of vibration is proportional to the square of the amplitude of vibration—twice the amplitude, four times the energy.

Since bats are not regular geometric objects, it is difficult to calculate the vibrational properties of a baseball bat. However, L. L. Van Zandt,[3] in one of the most elegant calculations in sports physics, has determined the vibrational modes of a standard wooden bat[d]. While the details of the calculations are not easily accessible, intelligent estimates of the amplitude of the vibrational effects can be made using *dimensional analysis.* The amplitude of the vibration induced by the collision of the bat and ball must be proportional to the impulse (the force applied to the bat multiplied by the time over which the force is applied), inversely proportional to the weight of the bat, and inversely proportional to the frequency of vibration. Knowing the impulse to the bat,

---

[3]An account of Van Zandt's work was published in the *American Journal of Physics* in 1992.

and the weight of the bat, and having measured the frequency of vibration of the bat, a typical amplitude of vibration induced in the bat by striking a ball near an antinode is determined to be a little less than one-half inch; the bat will be distorted about as much as the ball. Using the same kind of estimate, it seems that a significant fraction of the bat-ball collision energy can be transferred to the vibrational energy of the bat.

For the specific, though typical, bat considered here, the frequency of the oscillation was measured to be about 220 cycles per second (middle A). Since the corresponding characteristic half-cycle time of about 0.002 seconds is appreciably longer than the natural bat-ball interaction time of about 0.0005 seconds (see Figure 5.6), the bat does not return the energy of distortion to the ball but retains that energy in the vibration familiar to any baseball player. That energy is, of course, lost to the ball, hence, the ball will rebound with decreased velocity.

The solid curve of Figure 6.3 presents a rough estimate of the distance a batted ball travels as a function of the distance of the impact from the optimum point (which is near a node of the vibration) when the energy lost to the vibration is taken into account.

Although the half-period of vibration is sufficiently long compared to the time the ball is in contact with the bat so that the energy is not returned elastically to the ball, the act of deforming the bat does lengthen the time of bat-ball contact (as the bat gives way at the ball's impact). The highest-frequency sound produced in the course of ball-bat contact is roughly one-half the inverse of the impact time (so that a collision that lasts less than $\frac{1}{1000}$ of a second generates sound frequencies greater than 500 cycles per second, which is one octave above middle C). One hears these characteristic high frequencies in the "crack" of the bat. When the ball is hit off-center, the collision time is a little longer, generating somewhat lower maximum frequencies. With a lower maximum frequency and the addition of a strong component of lower-frequency sound from the natural bat oscillation, the "crack" becomes more of a "thunk." Outfielders use the sound as a part of the pattern to which they respond. A one-time center fielder told me that when the ball is hit straight at you (the distance such balls will travel is difficult to judge), if you hear the bat "crack," run back (toward the fence); if the sound is a "thunk," run in (toward second).

The curves of Figure 6.3 refer implicitly to a bat which retains its physical integrity through the collision. But we know that bats break, especially when the batter hits inside pitches near the handle. The phrase "sawed off the batter" is a metaphor with a bite. And when the batter is "sawed off" and delivers a broken-bat hit, the ball doesn't go very far. Certainly not as far as implied by Figure 6.3.

We can understand this reduction in the energy returned to the ball in the course of the collision by the breaking of the bat when we consider the forces exerted on the bat by the ball striking the bat near the handle. That collision bends the bat, stretching the wood fibers on the side of the bat away from the ball and compressing the fibers on the side near the ball. The response of those fibers results in an elastic resistance of the bat to the bending forces. The bending bat exerts a reaction force on the ball. As the bat bends under the ball-bat impact, the force acting on the ball increases as the bat bends; indeed, the force is roughly proportional to the amount of the bend. That force, directed away from the bat, is part of the force that propels the ball back toward the playing field. But when the wooden bat bends so much that the stress on the wooden fibers on the side of the bat away from the ball (toward the catcher) exceeds the strength of the stretched fibers, the fibers break and no longer resist the incursion of the ball.[4] With less force directed to returning the ball to the diamond, the velocity of the ball is much reduced and, instead of carrying over the infield, the ball sails feebly off of the broken bat into the infielder's glove.

It is interesting to note that the heavier, thick-handled bats of

---

[4]The tensile forces are greatest when the bending is large—away from the node. But shear forces, which tend to separate layers of fibers, are largest at the nodes where there is no bending. Both forces play a role in the breaking of the bat.

yesteryear were more rigid and had higher natural frequencies and smaller amplitudes of vibration upon off-center impacts with the ball; the wood fibers stretched less; hence, these bats broke less easily. Since hickory has about twice the elastic modulus of ash, the stiffer hickory bats also vibrated with a smaller amplitude and higher frequency. (Though, for a given configuration, the vibration frequency varies inversely with the square root of the density, and hickory is denser than ash, this reduction of the natural frequency from the greater density of hickory is not as important as the increase in frequency from the greater stiffness of the hickory.) Since the energy stored—and then lost—in the vibration of the bat is proportional to the square of the amplitude, the stiffer, though heavier, hickory bats will retain less of the collision energy for balls hit away from the sweet-spot.

Although the increase in bat mass of a thick-handled bat will slow the swing down somewhat, that effect is probably small. Hence, if a player is comfortable using a thick-handled bat, he might gain a couple of hits over a season on inside pitches, which now don't break his bat, and add a few points to his batting average.

Indeed, in general, bats with thick handles and bats with long barrels have longer vibration-free zones of good hitting and are broken less easily by inside pitches. Edd Roush, hitting choked up with his 48-ounce, thick-handled hickory bat in the 1920s, probably never in his life stung his hands hitting an inside pitch and seldom, if ever, broke a bat. The long-barreled "bottle bats" used by players such as Heinie Groh[5] and Bucky Dent make sense—for some players.

---

[5]Groh's bats had a barrel about 17 inches long which then necked down abruptly to a rather thin handle also about 17 inches long; this odd bat was about half-barrel and half-handle. Groh swung the 46-ounce bat choked up with his hands slightly apart. Groh had small hands and he preferred a thinner handle than was common in the 1920s. Dent used a more conventional bat but with a long full barrel which he swung choked up. Groh and Dent were largely line drive hitters, though both hit important home runs. Indeed, Groh led the 1919 Cincinnati Reds—who beat the Chicago Black Sox in the World Series—in home runs (with five) and in slugging average (at .431).

Recently, some have used the name "bottle-bat" to mean any thick-handled bat.

ABERRANT BATS, ALUMINUM BATS

Aluminum bats are now used in baseball played outside the major leagues. Bats constructed in strange and ingenious ways are used in baseball variants, such as various forms of recreational softball, where there seem to be no special constraints on bats. The use of illegally altered bats has occasionally been detected in major league baseball. Questions arise concerning the efficacy of these aberrant bats.

The thin-handled bats made of light wood that are popular today are fragile and not cheap. Hence, the cost of bats for amateur baseball and minor league baseball is significant on the economic scale of these activities. Since bats made from aluminum tubing do not break and can be made with appropriate balance and weight, they have become the bats of choice for such baseball and are now admitted by the rules. It is now hard to find a wooden bat among the rows of aluminum bats in most sporting goods stores.

The requirement that wooden bats be turned from one piece of solid wood places constraints on the size of light bats. Since woods less dense than ash are not strong enough to serve as material for bats, a light bat must perforce contain less wood and be smaller than a heavier bat. Moreover, the weight distribution along the bat is defined by the bat's shape, and the traditional shape is that which makes the best use of wood.

Since the thickness of the aluminum tubing used to make the hollow aluminum bats can be varied over wide limits, the weight and weight distributions of aluminum bats can be set almost independently of the size and shape of the bat. Presently, aluminum bats are shaped like traditional wooden bats and are made to have similar weights and weight distributions[e], but this follows from convention rather than as a mechanical necessity. The freedom of design afforded by aluminum has, however, resulted in the manufacture of light bats with the full-size 2.75-inch barrels allowed by the rules of baseball, and wooden bats must have smaller barrels to keep to the same weight. Hence, the typical 32-ounce wooden bat will have a barrel diameter of only 2.50 inches. With the larger barrel, the 32-ounce aluminum bat will have about a 4 percent larger effective bat-ball hitting surface with no compensating disadvantage.

Aluminum bats may also be better than wooden bats in other ways. The question has been raised as to whether a ball hit by an aluminum bat will go faster and farther than a ball hit by a wooden bat. Players also hold the view that an aluminum bat has a longer region of good hitting than a wooden bat. In particular, balls hit short on the bat—near the handle—are propelled more efficiently by an aluminum bat, and with less vibration and stinging of the hands, than by a wooden bat. There appears to be some consensus among players that these reputed advantages of the aluminum bat are real. Are their conclusions correct? If so, why?

We address first the question of the length balls are hit by an aluminum bat (or the speed with which ground balls hit by an aluminum bat travel through the infield). We have noted that the baseball is not very elastic. On striking a hard surface, only about 30 percent of the energy is returned in the rebound. The baseball is more elastic than a beanbag, but not by much. The coefficient of restitution for a beanbag striking a hard surface is near zero; beanbags do not bounce at all. No matter how strong the batter is, a baseball bat will not hit a beanbag much past second base. Nevertheless, it is possible to design an implement that will hit a beanbag a long way. An examination of such implements can provide some insight into mechanisms that allow baseballs to be hit farther by bats.

Though we will not be able to hit the beanbag very far with a baseball bat, we will do better with a tennis racket faced with a sheet of very thin rubber instead of gut or nylon strings. With this contraption, in the collision of the beanbag with the racket, the beanbag will travel past the plane of the racket by perhaps a foot, stretching the rubber. Then the rubber will react, propelling the beanbag away from the racket with a rather high velocity in a kind of catapult effect. Actually, a sheet of rubber would induce too much air resistance for a really fast swing of the racket-bat, and we would do better by stringing the racket with very strong, but quite elastic, strings.

If we attach a long handle to the tennis racket and weight the racket head to copy the length and weight distribution of a baseball bat, and then swing the racket like a bat, we will be able to hit a beanbag the size and weight of a baseball quite a long way. If we are able to find strings as elastic as gut or nylon that are sufficiently strong so they will not snap under the large forces we would generate, we could hit the beanbag farther than a baseball struck by a bat! (We could probably find steel springs that would do the job by replacing the conventional gut stringing of the racket.) Or we could hit a baseball farther with our racket-bat than with a regular wooden bat.

We explain this phenomenon by describing the (imaginary) observation of the results of a collision between a specific (also imaginary) racket-bat and the beanbag-ball (a baseball cover stuffed with beans so as to have the same weight as an official ball):

The beanbag was flung toward the plate by a major league fast-ball pitcher and the racket-bat was swung by a major league power hitter. When the beanbag hit the racket face, the strings gave way elastically and the bag moved 6 inches past the plane of the racket face. (This deflection is about ten times the compression of the baseball hit by a bat.) Then the strings reacted elastically, flinging the beanbag back toward the center-field bleachers.

The bat-beanbag collision took place in about $\frac{1}{100}$ of a second (rather than the $\frac{1}{1000}$ of a second for the collision of the wooden bat with the baseball), and the maximum force was 800 pounds (rather than the 8,000-pound force that reversed the baseball flight). The 800-pound force compressed the soft beanbag-ball one inch and distorted the racket-bat by 6 inches, as we noted. Then about six-sevenths of the collision energy was stored by the racket and only about one-seventh by the beanbag. Since the beanbag is completely inelastic, all the beanbag compression energy was lost to friction, but only 40 percent of the energy stored by the racket was lost. Then, for the collision of beanbag and racket, more than 50 percent of the collision energy was returned to the beanbag as kinetic energy in the catapultlike action of the racket on the beanbag, corresponding to a COR greater than 0.700. For a baseball struck by a wooden bat the energy return is only about 32 percent, corresponding to a COR of about 0.565. A racket-bat swing, which would drive a baseball 400 feet using a wooden bat, would drive the beanbag about 480 feet (and a baseball a few feet farther). The racket-bat would hit a beanbag farther than the wooden bat would hit a baseball! And if the racket-bat were used on a baseball in Yankee Stadium, the ball could well be knocked clear out of the stadium—and not even Babe Ruth ever hit a ball out of Yankee Stadium.

Even an umpire as obtuse as fans claim in their calumny would be suspicious of such a bat. But aluminum bats seem to have a little of the catapultlike elastic properties of the racket-bat and probably can hit a baseball appreciably farther than a wooden bat. From measurements made by my colleague R. C. Larsen on the compressibility of aluminum bats, it seems that, for a given force, the distortion of an aluminum bat is about one-tenth ^{f} as great as the distortion of the ball (rather than the factor one-fiftieth for wooden bats). Moreover, from the high frequency and persistence of the sound emitted by the bat when struck lightly at a hitting point (at a node of the longitudinal vibration), it appeared that the distortion was quite elastic and the compression energy would be restored quickly (unlike the longitudinal vibration). Hence, in this mode the bat was surely much more elastic than the ball while the relatively incompressible wooden bat stores little energy and is only slightly more elastic than the ball. Then, the aluminum bat stores about one-eleventh of the collision energy in a highly elastic deformation of the bat, which was returned efficiently to the ball, and the ball stored about ten-elevenths of the energy in the deformation of the ball, most of which was lost in friction.

Though the intrinsic elasticity of the bat will be near 100 percent, some of the bat deformation energy will be retained in the energy of motion of the aluminum shell that makes up the bat. We estimate that 80 percent may be returned to the ball.

The elasticity of the ball at this impact velocity will be about 19 percent, which is the square of the COR at the relevant impact force. Using these values, 24.5 percent of the collision energy will be restored to the bat and ball, a value much greater than the 19 percent for collision of the wooden bat and ball. This corresponds to an effective COR of 0.49 rather than the value of 0.44 appropriate for the wooden bat-ball impact at that impact velocity, and the home run hit 380 feet with the wooden bat will go as much as 410 feet from the aluminum bat. The same kind of calculations suggest that a fungo hit 330 feet by a (regular) wooden bat would travel about 12 feet farther hit with the same swing from a similar aluminum bat.

Though the numerical values presented here can be considered only as a kind of illustrative estimate—in view of our uncertain understanding of the character of the aluminum bat—the aluminum bats now used for baseball will surely hit a ball appreciably farther than a wooden bat.

The second purported advantage of the aluminum bat is that it is more effective than wood on inside pitches hit too near the handle. The aluminum bat is a kind of shaped hollow cylinder; since aluminum is about four times as dense as ash, to achieve the same weight the bat must be hollow. Aside from the greater strength of the aluminum, the hollow cylinder is more rigid than a solid structure containing the same mass of material; typically, the aluminum bat is about twice as stiff as the wooden bat—twice as much force is required to bend it a given amount. Hence, when a ball is struck badly—near an antinode at the handle or the end of the bat—the stiffer aluminum bat takes up much less energy in longitudinal vibration than the wooden bat—and stings the hands less. Moreover, since the vibrational frequency is higher than for a wooden bat, the aluminum bat even returns some of that energy to the ball. Hence, the response of the aluminum bat to a ball hit away from the vibrational node (or sweet-spot) is close to the dotted curve of the perfectly rigid bat shown in Figure 6.3. A ball mis-hit, near the handle or the end of the bat, will go farther off an aluminum bat than a wooden bat. There is more room for error with an aluminum bat.

Perhaps more important, the aluminum bat does not break when a ball is hit near the handle. The force of resistance to the ball supplied in the course of the bending of the bat is not truncated by the breaking of the bat. With more force propelling the ball back, the ball will come off of the bat faster—perhaps over the infielder's head—than from the broken wooden bat.

ABERRANT BATS, CORKED BATS

On occasion, players have modified their wooden bats by drilling an axial hole in the end of the bat and filling it with elements such as cork, cork balls, and rubber. They place a wooden cap over the hole and sand and varnish it to hide the modification. Neither the hole nor the addition of the filling are allowed under official baseball rules.

It is desirable to determine whether such an illegal modification might create a bat with properties that a legal bat cannot achieve—and hence threaten the integrity of the game—or if the effect of the modification can only produce a bat with properties no different than might be achieved with a legal bat design.

It is not possible to conclude categorically that a specific physical change in a bat is good or bad. But it is possible to determine the character of the change and to determine if the change—if illegal—would produce a bat with characteristics outside the range of bats constructed according to baseball rules.

The properties of a bat relevant to striking a ball squarely are largely defined by three weight distributions, or three moments. We list these together with representative values (taken from a particular bat that was 35 inches long and weighed 32 ounces). All weights are expressed here in ounces, and lengths are expressed in inches.

a. The sum of the weight of each piece of the bat, which is just the total weight of the bat (the zero moment) [32 ounces].
b. The sum of the weight times distance, measured from the handle, for each piece of the bat (the first moment) [765 inch ounces].
c. The sum of the weight times the square of this distance for each piece of the bat (the second moment or moment of inertia) [21,000 square-inch ounces].

There are three key positions along the bat that follow from the moments—defined here as lengths measured from the handle end of the bat.

d. The center of gravity [24 inches].
e. The center of inertia [25.6 inches].
f. The center of percussion (a sweet-spot) [27.3 inches].

There are other factors of interest:

g. The liveliness or elasticity of the bat.
h. The resonant frequency of the bat [220 cycles per second].
i. The position of the vibrational node [27.35 inches].

Though the nodal point and center of percussion will be close for most bats, the near equivalence for this particular bat is an accident.

For a given bat, the center of gravity (d) is just the point of balance of the bat—as balanced on a finger, for example. The center of percussion (f) can be found by holding the bat lightly at the handle end, e.g., by finger and thumb, and striking the barrel with a light hammer. When the bat is struck at the center of percussion, one feels no motion at the handle. For most bats, this point is very near the vibrational node (i); when the bat is struck by the hammer at the node, no vibrations will be felt at the handle end. The center of inertia (e) can also be determined primitively but less simply. If the bat is laid out on a smooth, waxed wooden floor (or better, on an ice surface) and struck or pushed vigorously at a point so that the bat begins to slide without rotating appreciably (in the plane of the surface), that point is the center of inertia.

The moments are physically manifest in straightforward ways. The weight (a) is felt when one simply holds the bat vertically. The force required to hold the stationary bat out horizontally is proportional to the first moment (b). The force required to wave it back and forth vigorously when it is vertical is proportional to the second moment (c). This second moment contributes most to the "feel" of the bat and is the factor most important to the batter.

The elasticity (g) is determined by the bat's resilience near the point of the bat that strikes the ball; a resilient bat may store energy upon impact and return that energy to the ball. Wooden bats are not very resilient—and thus store little energy. The resonant frequency (h) is a measure of the energy loss when a ball is struck away from the vibrational node. A high frequency indicates a larger, i.e., longer, sweet-spot. Bats with long barrels or thick handles will display higher vibrational frequencies and larger sweet-spots.

An interesting illegal bat modification is made by drilling an axial hole in the end of a bat and filling it with an extraneous material (e.g., cork, cork balls, rubber, rubber balls). We first consider the effects of the hole—without filling.

As an exercise, I have been able to show explicitly that for a specific illegal modification of a specific bat, the changes in the moments (a, b, and c) and the characteristic lengths (d, e, and f) that define the bat's character can be achieved through legal changes. The changes will not affect the elasticity (g). While it is possible to invent extreme illegal corklike modifications that could not be copied legally, e.g., making the whole end of the bat cork, such bats would probably not be much good in hitting a baseball.

We compared the characteristics of a specific bat before and after a specific illegal modification and after certain legal modifications. The bat was an inexpensive "Louisville Slugger–Wade

Boggs Model" purchased from a local sporting goods store. It was 35 inches long and weighed 32 ounces. The illegal modification was chosen to be a hole $1\frac{3}{16}$ inches in diameter drilled 6 inches deep along the axis from the barrel end of the bat. The hole was filled with cork. The density of the wood was 0.638—a typical density for American ash (or white ash)—and the density of the cork was 0.25 times the density of water.

Such a modification does sensibly change the bat. The weight is reduced by about 1.5 ounces, and that loss is wholly at the end of the bat. If a player is having trouble getting around on the fast ball, this weight reduction will help him in very much the same way that a lighter bat might help: the batter will probably not drive the fast ball as far by 2 or 3 feet when he hits it well with the drilled-out bat, but he may hit it with good timing more often.

Aside from eliminating the hollow sound from the air column created by the hole in the bat (such a sound might excite an excess of interest by the umpire), the cork or rubber stuffed into a hole drilled in the bat will serve more as a detriment than an advantage. The extra material will add, perhaps, one-half ounce to the bat's weight—at the end—and then store about 2 percent of the bat's energy. But that energy will not be effectively transferred to the ball. Even if the filler were quite elastic, the elastic energy of the filler cannot be transferred efficiently to the bat in the $\frac{1}{1000}$ second of the bat-ball collision. Hence, the extra material will only slow the bat down a little and slightly reduce the distance the bat hits the ball. Such a filler would thus take about 3 feet more off a 100-foot drive.

Whatever the advantages of the lighter bat, it was possible to show that very near the same results achieved by the illegal modifications could be achieved legally by (a) choking up on the bat by about one inch, (b) using a bat with the same dimensions but constructed of slightly lighter wood, (c) sawing (legally) about $\frac{3}{4}$ of an inch off the end of the original bat, and (d) legally turning the barrel down from 2.50 inches to 2.40 inches. Both the modified bat (d) with the reduced barrel size and a bat shortened by one-half inch and made of slightly lighter wood fit the moments and characteristic lengths of the illegally drilled bat almost exactly.

Solution (b) implies that a satisfactory "lighter" wood is available, which may not be the case. The American ash from which bats are made has an unusually high strength-to-weight ratio. Ash was celebrated in medieval times as the only proper wood from which to construct the lances of knights errant; an ash lance was light enough to carry and wield and strong enough to impale the opposition. Any reduction in the weight of the wood seems to require a substantial compromise in strength and hardness. But green (or Pennsylvania) ash or black walnut, for example, might provide adequate lighter—if somewhat weaker—bats.

We have noted that aluminum bats are much more resilient than wooden bats and upon striking a ball store energy in compression, which is returned to the ball efficiently. It is conceivable that a device could be fashioned to insert into a hole drilled in a bat that could store energy upon impact, which would then be returned efficiently to the ball (in much the same way that the flexion of an aluminum bat stores, and returns, energy). Such a device could improve the effectiveness of a bat, but some reflection suggests that it would be difficult to design—and quite difficult, if not impossible, to make so that it would be hard to detect.

The resonant frequency and position of the vibrational node are only important for miss-hit balls. Nevertheless, such illegal holes drilled in the bat have no significant effect on either that frequency or the node's position.

In summary: The characteristics of any specific baseball bat can be changed by drilling an axial hole in the end of the bat and filling it with some light, inactive, extraneous material. The modified bat differs from the original bat by its lighter weight and smaller moment of inertia. Bats drilled out in this way are excluded from play under the *Official Baseball Rules*. But the properties of such modified bats can generally be reproduced by a legally constructed bat. In particular, a slightly shorter bat made of the same-density wood, or a bat with a slightly smaller barrel, or the same-size bat made of somewhat less dense wood will have nearly the same "feel" and hitting characteristics as the bat lightened by having a section of the core replaced by a lighter material.

The only substantial caveat to this conclusion follows from the limitations on properties of wood. If the batter demands a lighter bat *exactly* the size of his regular bat—and will not tolerate a modest reduction in hardness and strength—he can best achieve that result by illegally drilling a hole in it.

TECHNICAL NOTES

a. In this model, we calculate the maximum velocity v of the bat, from the relation

$$v = \sqrt{\frac{2E}{m + \epsilon^2 M}}$$

where m is the mass of the bat, M the mass of the player, and E is a normalization constant approximately equal to the total energy produced by the muscles. The constant, ϵ, is taken as equal to $\frac{1}{5}$, which is in accord with the observation that the maximum energy of bat and player are roughly equal. Using this relation, for a given total energy E, we determine the variation of v with m for batters of a given weight M. Then, as discussed earlier, we determine the velocity that a pitched ball will travel when struck by the bat of mass m moving with the velocity v.

In the approximation that the ratio of muscle mass to total mass is about the same for players of different size, we can expect that the energy put out by the muscles is proportional to the weight of the player. (For example, if the shape of a small player were the same as a player 30 percent heavier, the muscles of the larger player, 10 percent longer and with a 20 percent larger cross section, would be expected to exert a 20 percent greater force over a distance 10 percent greater to develop 30 percent more energy.) Hence, we can set E proportional to M and estimate how much farther a large player might be expected to hit a ball than a smaller player.

b. The approximation that the bat can be treated as a free piece of lumber, which is adequate for balls hit by the bat near the center of percussion (effectively the sweet-spot), may break down for balls hit too near the handle or the end of the bat. The clamping action of the hands restricts the rotation of a (unrealistically) stiff bat reducing the energy loss to such rotation. Hence, the broken curve must overestimate the energy loss of a stiff bat—if anyone can find such a bat—for impact points far from the sweet-spot. This suggests that a firm grip on the bat might help for balls hit on the end or the handle of the bat—while a loose grip would do as well for a well-hit ball. But that advantage of a firm grip is reduced if the bat is as flexible as most wooden bats are. For a real bat, the signal of the impact travels along the bat as a transverse vibrational wave at a velocity that varies along the bat—very fast at the thick barrel, more slowly near the thin handle. We can estimate the velocity of the signal by assuming that the typical 20 inch ($1\frac{2}{3}$ feet) distance from the node at the barrel and the antinode at the handle is about one-half wavelength. Using the measured vibrational frequency of 220 cps, the signal speed will then be about $2 \cdot (1\frac{2}{3}) \cdot 220 = 733$ ft/sec and the signal will take about 2 milliseconds to reach the hands from the point of bat-ball impact. With the duration of that impact of the order of $\frac{1}{2000}$ of a second, the handle and hands will not receive the signal—and respond to the impact—until the ball has left the bat.

c. The position of the nodes of the fundamental vibration of a wooden bat can be easily determined with no more than the bat, a piece of string, and a light hammer. With the addition of a piano and a sense of pitch, one can also find the frequency. Suspend the bat by the string tied around the knob, tap the bat with the hammer, and listen. When the bat is struck at the end or where the barrel necks down to the handle, the tap will cause the bat to ring for a fraction of a second. As you move the position of the tap, you will find two points where there is no such ring, one point in the region of the trademark, one 6″ or so from the handle end. These are the vibrational nodes. By matching the bat ringing frequency with notes struck on the piano, you should be able to determine the approximate frequency of the bat, which will probably be near A below middle C or 220 Hertz.

This procedure does not work well for an aluminum bat, as the striking of the bat anywhere incites a vibration of the column of air in the hollow bat. This organ pipe–like air-column vibration will be near the same frequency as that of the fundamental bat vibration and will obscure it.

There are higher vibrational frequencies with three, four, and more nodes, but these are not so important.

When the bat (wood or aluminum) hanging from two feet of string tied about the handle is struck a little harder well above the trademark, the handle end of the bat will jump in the direction of the hammer blow. But if the bat is struck on the end of the bat, the handle will jump in the opposite direction. There will be a point between where the handle does not react at all when the bat is struck. This is the *center of percussion.* For a baseball bat, the center of percussion is usually very close to the vibrational node and in the vicinity of the trademark.

For any bat, the region containing the vibrational mode and the center of percussion is called the "sweet-spot." When the ball is hit at the sweet-spot, there is no vibration and no recoil on the hands, the hit feels good, and the ball shoots off of the bat.

d. Most calculations of physical quantities require some level of approximation. The mathematical model is almost never a precise map of reality. Nevertheless, Professor Van Zandt's calculations of the characteristic vibrational frequencies and amplitude shapes for a baseball bat should be accurate; however, his calculated frequencies were about 25 percent lower than measurements of those frequencies by Prof. Uwe Hansen. With that correction (which might have followed from a difference in the elastic properties of wood chosen for bats), the calculated frequency of the fundamental bat vibration and twenty harmonics fit the measured frequencies within about one percent. The fundamental frequency of the bat Van Zandt considered was 170 Hz; the first harmonic was 560 Hz.

e. In fact, manufacturers seem to make aluminum bats with somewhat different weight distributions than wooden bats. Since the weight of a section of the hollow aluminum bat is in the shell or skin of the bat while the mass of the similar section of the solid wooden bat is distributed through the whole section, there is a tendency to place relatively less weight in the barrel of the aluminum bat and more in the handle. If the same gauge (or thickness) of aluminum were used throughout the bat, a one-inch section of a 2.75″ diameter barrel would weigh about 2.75 times as much as a one-inch section of the handle. For a wooden bat of the same dimensions, the ratio would be much greater—about $(2.75/1.0)^2 = 7.56$. Usually, the manufacturer only partially counters this by using thicker material in the barrel than in the handle of the aluminum bat.

The freedom of design of aluminum bats has the consequence that nominally similar bats may have quite different properties. Hence, the comments I make concerning aluminum bats may not apply to all bats.

The different weight distribution of the typical aluminum bat has both advantages and disadvantages. With less weight in the barrel behind the impact with the ball, the 32-ounce aluminum bat acts in the collision with the ball more like a 30-ounce wooden bat, transfers less energy to the ball, and some distance is lost on a solid hit. But the trampoline action of the bat acts to drive the ball much farther than a wooden bat anyway and the effectively lighter bat can be swung a little quicker. Also, since the extra weight in the handle increases the turning moment of the bat, a ball mis-hit too near the end of the bat or the handle might lose a little less energy in turning the bat if the bat were quite stiff and loosely held by the batter. But this advantage is, at best, small and the reduction of vibration by the stronger handle of the aluminum bat is more important in increasing the length of the good hitting region.

f. The measurements of the compressibility of the balls and bats were made statically and can only approximate the actual forces generated by the collision of the spherical ball with the cylindrical bat. We believe the results more likely underestimate, than overestimate, the distortions of the bats.

INDEX